Communications in Computer and Information Science   1978

Rationale

The CCIS series is devoted to the publication of proceedings of computer science conferences. Its aim is to efficiently disseminate original research results in informatics in printed and electronic form. While the focus is on publication of peer-reviewed full papers presenting mature work, inclusion of reviewed short papers reporting on work in progress is welcome, too. Besides globally relevant meetings with internationally representative program committees guaranteeing a strict peer-reviewing and paper selection process, conferences run by societies or of high regional or national relevance are also considered for publication.

Topics

The topical scope of CCIS spans the entire spectrum of informatics ranging from foundational topics in the theory of computing to information and communications science and technology and a broad variety of interdisciplinary application fields.

Information for Volume Editors and Authors

Publication in CCIS is free of charge. No royalties are paid, however, we offer registered conference participants temporary free access to the online version of the conference proceedings on SpringerLink (http://link.springer.com) by means of an http referrer from the conference website and/or a number of complimentary printed copies, as specified in the official acceptance email of the event.

CCIS proceedings can be published in time for distribution at conferences or as post-proceedings, and delivered in the form of printed books and/or electronically as USBs and/or e-content licenses for accessing proceedings at SpringerLink. Furthermore, CCIS proceedings are included in the CCIS electronic book series hosted in the SpringerLink digital library at http://link.springer.com/bookseries/7899. Conferences publishing in CCIS are allowed to use Online Conference Service (OCS) for managing the whole proceedings lifecycle (from submission and reviewing to preparing for publication) free of charge.

Publication process

The language of publication is exclusively English. Authors publishing in CCIS have to sign the Springer CCIS copyright transfer form, however, they are free to use their material published in CCIS for substantially changed, more elaborate subsequent publications elsewhere. For the preparation of the camera-ready papers/files, authors have to strictly adhere to the Springer CCIS Authors' Instructions and are strongly encouraged to use the CCIS LaTeX style files or templates.

Abstracting/Indexing

CCIS is abstracted/indexed in DBLP, Google Scholar, EI-Compendex, Mathematical Reviews, SCImago, Scopus. CCIS volumes are also submitted for the inclusion in ISI Proceedings.

How to start

To start the evaluation of your proposal for inclusion in the CCIS series, please send an e-mail to ccis@springer.com.

Raja Muthalagu · Tamizharasan P S ·
Pranav M. Pawar · Elakkiya R ·
Neeli Rashmi Prasad · Michele Fiorentino
Editors

Computational Intelligence and Network Systems

First International Conference, CINS 2023
Dubai, United Arab Emirates, October 18–20, 2023
Proceedings

Springer

Editors
Raja Muthalagu 🆔
Birla Institute of Technology and Science
Dubai, United Arab Emirates

Pranav M. Pawar 🆔
Birla Institute of Technology and Science
Dubai, United Arab Emirates

Neeli Rashmi Prasad 🆔
TurstMobi
San Francisco, USA

Tamizharasan P S 🆔
Birla Institute of Technology and Science
Dubai, United Arab Emirates

Elakkiya R 🆔
Birla Institute of Technology and Science
Dubai, United Arab Emirates

Michele Fiorentino 🆔
Polytechnic University of Bari
Bari, Italy

ISSN 1865-0929   ISSN 1865-0937 (electronic)
Communications in Computer and Information Science
ISBN 978-3-031-48983-9   ISBN 978-3-031-48984-6 (eBook)
https://doi.org/10.1007/978-3-031-48984-6

This Springer imprint is published by the registered company Springer Nature Switzerland AG
The registered company address is: Gewerbestrasse 11, 6330 Cham, Switzerland

Paper in this product is recyclable.

Preface

We proudly present the post-conference proceedings of the International Conference on Computational Intelligence and Network Systems (CINS 2023). The inaugural CINS 2023 conference took place at the Birla Institute of Technology and Science Pilani, Dubai Campus, located in Dubai, United Arab Emirates. The conference is occurred from October 18th to October 20th, 2023. The primary objective of this conference was to facilitate the development of "Seamless Computing for the Next Generation." It provided a venue for the convergence of academia and industry, and forums to collectively investigate the domains of computing, intelligent approaches, and networks. The conference facilitated the promotion and exploration of novel and ambitious ideas, trends, and future problems pertaining to seamless computing. The subject matter encompassed contemporary challenges within computing systems and the utilization of intelligent approaches to improve computing methodologies, data processing capabilities, and the application of said intelligent techniques. The conference also addressed several topics pertaining to networks, including security, network data processing, networks that transcend boundaries, device heterogeneity, and advancements in networks connected to the Internet of Things, software-defined networks, cloud computing, and intelligent networks, among others.

The conference received full papers from four distinct tracks, including computational intelligence, data engineering, security and privacy, and intelligent network technologies. A total of 130 complete paper submissions were received. All manuscripts included in the planned proceedings were submitted through an online platform and underwent a peer-review process. The review process used a single-blind approach, where the identities of the reviewers were kept anonymous while the authors' identities were known. The reviewers were selected from a group of experienced program committee members and each submission was assessed by three independent reviewers. A total of 11 full papers, accounting for 8% of the submissions, have been deemed suitable for inclusion in the present proceedings volume of the Springer CCIS Series. The authors had the chance to revise the paper in response to helpful criticism from the expert reviewers, and they then submitted an updated, camera-ready version of the manuscript.

All accepted submissions were presented during the conference as a 15-min presentation talk and a 5-min question-answer from session chairs and the audience. The conference was attended by leading industry people and academicians. The conference keynote sessions were delivered by Christoph Benzmüller, Chair for AI Systems Engineering, Otto-Friedrich-Universität Bamberg & FU Berlin, Germany, and Jyotika Singh, Director of Data Science, Placemakr, USA. We also organized a workshop session on Intel One API and Cyber and Digital Forensics Workshop by Shriram K. Vasudevan, Intel Software Innovator, Intel Corporation Ltd., USA, and Nikhil Mahadeshwar, Founder, Cyber Secured India, India. In addition to scientific presentations, expert keynote sessions, and workshops, the other major attraction of the conference was the Microsoft ESL Generative AI Global Summit. The Microsoft ESL Generative AI Global Summit presented interesting ideas from different people around the globe on an AI-driven system that utilizes generative AI techniques to accurately and expressively convert written or spoken Arabic text into Emirati Sign Language (ESL) animations or descriptions. In conclusion, we express our sincere gratitude to each person who made this conference and all events at the conference successful, which includes Birla Institute of Technology and Science (BITS) Pilani, Dubai Campus, Dubai, UAE, the local organizing committee, volunteers, program committee members, session chairs, all expert reviewers, and the proceedings publishers. Finally, yet most important, we truly acknowledge our authors, who contributed their scientific knowledge to the conference and community. We sincerely hope that this volume of proceedings will give readers a spark of knowledge and innovative ideas.

Raja Muthalagu
Tamizharasan P S
Pranav M. Pawar
Elakkiya R
Neeli Rashmi Prasad
Michele Fiorentino

Organising Team

Chief Patron

V. Ramgopal Rao
  (Vice Chancellor)
BITS Pilani, India

Patron

Srinivasan Madapusi (Director)   BITS Pilani, Dubai Campus, Dubai, UAE

Program Chair

Vijayakumar B.   BITS Pilani, Dubai Campus, Dubai, UAE

Program Co-chairs

Raja Muthalagu   BITS Pilani, Dubai Campus, Dubai, UAE
Tamizharasan P S   BITS Pilani, Dubai Campus, Dubai, UAE
Pranav M. Pawar   BITS Pilani, Dubai Campus, Dubai, UAE
Elakkiya R   BITS Pilani, Dubai Campus, Dubai, UAE
Neeli Rashmi Prasad   TurstMobi, USA and SmartAvatar,
                          The Netherlands
Michele Fiorentino   Polytechnico di Bari, Italy

Organizing Secretaries

Raja Muthalagu   BITS Pilani, Dubai Campus, Dubai, UAE
Tamizharasan P S   BITS Pilani, Dubai Campus, Dubai, UAE
Pranav M. Pawar   BITS Pilani, Dubai Campus, Dubai, UAE
Elakkiya R   BITS Pilani, Dubai Campus, Dubai, UAE

Local Organizing Committee

Vadivel S.	BITS Pilani, Dubai Campus, Dubai, UAE
Sujala D. Shetty	BITS Pilani, Dubai Campus, Dubai, UAE
Siddhaling Urolagin	BITS Pilani, Dubai Campus, Dubai, UAE
Angel Arul Jothi	BITS Pilani, Dubai Campus, Dubai, UAE
Pramod Gaur	BITS Pilani, Dubai Campus, Dubai, UAE
Ashish Gupta	BITS Pilani, Dubai Campus, Dubai, UAE
Sapna Sadhwani	BITS Pilani, Dubai Campus, Dubai, UAE
Ram Krishna Mishra	BITS Pilani, Dubai Campus, Dubai, UAE
Kashyap Shah	BITS Pilani, Dubai Campus, Dubai, UAE
Sheeba Uruj	BITS Pilani, Dubai Campus, Dubai, UAE

Program Committee

Richard Ikuesan	Zayed University, UAE
Zakaria Maamar	Zayed University, UAE
Monther Aldwairi	Zayed University, UAE
Ravishankar Sharma	Zayed University, UAE
R. Rama	IIT-Madras, India
Gerassimos Barlas	American University of Sharjah, UAE
Imran A. Zualkernan	American University of Sharjah, UAE
Saad Harous	University of Sharjah, UAE
Praveen Kumar Yadav	Panasonic R&D Centre, Singapore
Wathiq Mansoor	University of Dubai, UAE
Emad Bataineh	Zayed University, UAE
Muhammad Shafique	NYU Abu Dhabi, UAE
Rasmus H. Nielsen	Movimento, USA
Siddhartha Bhattacharyya	Florida Institute of Technology, USA
Hesham El Sayed	United Arab Emirates University, UAE
Rahat Iqbal	University of Dubai, UAE
Sami Miniaoui	University of Dubai, UAE
Vinod Pangracious	American University of Dubai, UAE
Prabhat Kumar Upadhyay	IIT-Indore, India
Alavi Kunhu	University of Dubai, UAE
Sandeep Jadhav	Dubai Holding, UAE
Rahul C. S.	IIT Goa, India
P. Muthuchidambaranathan	NIT Tiruchirappalli, India

N. Ramasubramanian	NIT Tiruchirappalli, India
Dnyaeshwar Mantri	University of Pune, India
Maha Saadeh	Middlesex University Dubai, UAE
Antonis Michalis	Tampere University, Finland
Samia Loucif	Zayed University, UAE

Contents

SemVidRec: A Semantic Approach to Annotations Driven Video Recommendation Model Incorporating Machine Intelligence

Arya Adesh[1], Gerard Deepak[2(✉)], and A. Santhanavijayan[3]

[1] Department of Computer Science Engineering, RV College of Engineering, Bengaluru, India
[2] Department of Computer Science Engineering, Manipal Institute of Technology Bengaluru, Manipal Academy of Higher Education, Manipal, India
gerard.deepak.christuni@gmail.com
[3] Department of Computer Science Engineering, National Institute of Technology, Tiruchirappalli, India

Abstract. Video recommendation ensures that viewers get content more relevant to their choices and taste. With the aggregation of a diverse variety of content on video streaming platforms, there is a need to improve the existing recommendation model to increase accuracy in predicting the best content analogous to past choices of the user. The model proposed in this paper employs semantic similarity to generate recommendations based on metadata and annotations. The semantic inference-based model ensures the incorporation of cognitive knowledge of user inputs and past activities, yielding better insights into user preferences. The model utilizes web-sourced metadata to augment its dataset with supplementary information. The metadata is classified using RNN and is stored in a semantic network for feature selection. Topic modeling on query terms allows the discovery of hidden topics in the user query and the use of a knowledge store like DBpedia allows for enriching the query terms based on the topics discovered. The video dataset is categorized using a combination of Random Forest and decision tree-bagging techniques, which are based on the extracted features of the videos. The proposed SemVidRec model is compared with other video recommendation models like VAVR, PVRRC, CCVR, etc. SemVidRec was discovered to outperform the baseline models based on several performance metrics, achieving an accuracy of 96.45%, precision of 95.43%, and an nDCG value of 0.97.

Keywords: Bagging · Semantic Similarity · Semantic Network · DBpedia · Structural Topic Modeling

1 Introduction

Video recommendations are video-content suggestions made to the user on a video streaming platform, narrowed down to best suit the needs and taste of that user. Video recommendations help filter out videos most likely to be watched, hence separating irrelevant content. This builds up a productive environment for the user, by reducing

© The Author(s), under exclusive license to Springer Nature Switzerland AG 2024
R. Muthalagu et al. (Eds.): CINS 2023, CCIS 1978, pp. 1–13, 2024.
https://doi.org/10.1007/978-3-031-48984-6_1

time spent on searching and selecting relevant content. Video recommendation systems draw inferences about user preferences based on watch history or search queries. With a rise in popularity and content in video aggregation and streaming platforms, for instance, YouTube and OTTs like Hulu, Amazon Prime, Netflix, and others, there is a need for an efficient and improvised video recommendation model. The two most frequently used recommendation models are content-based filtering and collaborative filtering.

Content-based recommendation model extracts visual features from individual frames of video and suggests content by investigating and matching these features with other videos. Collaborative filtering is a recommended approach that utilizes data obtained from multiple users to suggest items based on the similarity of users' preferences. The hybrid model for recommendation is a combination of the aforementioned video recommendation models, which either cumulatively adds abilities from both models or by merging their separate results.

However, these video recommendation models are not feasible to make predictions, when the video content on the platform is huge. Hence there is a need for a recommendation model that takes in auxiliary data and cognitive insights to make predictions accurately. Annotations-based video recommendation model uses video labels as the basis for prediction. This model can handle huge video content more effectively than content-based or collaborative filtering models. It is also important to make the recommendation model compatible with Semantic web standards. The semantic web has all content organized in machine-understandable form and hence allows the encoding of semantics with data. The inclusion of semantic insight into the recommendation model helps learn about user preferences, better, yielding accurate predictions. In the semantic web, metadata describing the content available in the World Wide Web is machine-understandable. This metadata can be used by the model to include data semantically similar to the query terms, allowing the inclusion of auxiliary information. The model proposed in the paper is completely compatible with Web 3.0 hence more relevant to the real-world scenario.

Motivation: The World Wide Web is extending into the Semantic Web (Web 3.0). Data on the semantic web is linked and exists in the form of a knowledge graph to facilitate machine understandability. This makes Web 3.0 powerful, robust, and agile. The existing recommendation model may not work competently on such an organized web of content, as they are based on older learning paradigms. There is a need for a recommendation model which includes semantic insights for making accurate predictions to improve user engagement. In this study, a video recommendation model is proposed that incorporates machine intelligence, which is a combination of machine learning and artificial intelligence paradigms. This model is designed to cater to the requirements of the Semantic Web. The video recommendation system named SemVidRec, is an ontology-based model to be able to process cohesive and huge data on the web. This improvisation helps.

Contribution: The proposed method introduces several novel contributions. Firstly, a semantic network is derived, which is based on enriched queries from the user history and insights extracted from the user profile. Shannon's entropy is utilized to construct the semantic network based on metadata instances classified by a strong deep-learning model. Secondly, the method encompasses Structural topic modeling and harvests entities from the DBpedia knowledge store to ensure the relevance and diversity of query terms. Thirdly, a dataset is classified by extracting features from the semantic network by using strong bagging techniques. Fourthly, the method employs semantic similarity models with empirically decided differential thresholds to rank and make relevant recommendations. Finally, testing confirmed a higher%age of precision, recall, F-Measure, and accuracy while ensuring a very low False Discovery Rate (FDR).

Organization: The paper is structured as follows: Sect. 2 explores Relevant Work, Sect. 3 provides an overview of the Planned System Architecture and details the implementation of the SemVidRec model. Section 4 highlights the Findings and Evaluation of the model's performance. This paper concludes in Sect. 5.

2 Related Work

Elahi et al. [1] proposes a model to solve the moderate cold-start problem as it extracts visual features from individual frames of a video using color histogram distance. Features extracted from the individual frames are aggregated using aggregation functions and recommendation.

Is done based on the K-Nearest Neighbor algorithm. Cosine similarity is used on the nearest neighbor set and the predicted preference score is calculated. Du et al. [2] proposes a model that uses both textual and non-textual features from the videos which are fused with priority-based late fusion. Collaborative embedded regression was used to overcome the drawback of the unavailability of one specific content. The proposed model is designed to operate in both in-matrix and out-of-matrix scenarios.

Zu et al. [3] Proposed a model to find suitable resources for online learning. The model constructs a subgraph on a seed set of recommendations. By constructing a cross-curriculum video-associated knowledge map and applying a random walk algorithm, the model recommends relevant subgraphs of course videos to learners. Vimala et al. [4] proposes a Kullback-Leibler divergence-based fuzzy C-means clustering method with improved square root cosine similarity to enhance the accuracy of collaborative filtering movie recommendation systems. To enhance the efficiency of the suggested approach, Support Vector Machines (SVM) are utilized to achieve more precise predictions. For comparison Fuzzy+ SVM+ Cosine model which is an improvisation of the model proposed by Vimala et al. [4], is used as one of the baseline models.

Yan et al. [5] proposes a hybrid collaborative filtering system based on a deep autoencoder model which intends to solve cold-start recommendation problems based on users' viewing behavior. The user/item information is processed by data processing and embedded layers, the auto-encoder layer is responsible for mining implicit features and the correlation between them. The final prediction is done by mapping the target rating vector with the feature matrix in the multi-layer perceptron layer. Cai et al. [6] presents an approach to active learning for video recommendation, which uses a multi-view strategy that leverages the visual characteristics of a video while requiring few annotations. A visual-to-text mapping function is utilized to map visual features and textual views to minimize classification loss. The model uses watching frequency and prediction inconsistency to select videos for metadata querying. Ma et al. [7] put forward a model for suggesting micro-videos available on social networks. This model uses user-item inter-action features, textual features, visual features extracted from videos. Using a deep neural network-based latent genre learning, it is possible to identify the concealed genres of micro-videos. This approach is proposed to enhance the recommendation quality by recognizing hidden patterns within the content, enabling more precise suggestions for the viewer.

Bhatt et al. [8] put forward a content-based recommendation model for online course content that uses sequential pattern mining of inter-topic relationships. This approach is proposed to create a personalized learning experience and to suggest courses that are relevant to the individual's interests based on their past interactions. The inter-topic relationships are mined from the instructor's syllabi as they tend to exist in educational corpora. The prediction is done based on content similarity between videos determined by Topic Similarity Score, Global, and Local sequence scores. Mei et al. [9] suggests an innovative recommendation system for online videos, which utilizes a collection of relevant videos based on their multimodal relevance (written, graphic, and audio components) and user clicks. A relevance mechanism is used to find and assign optimum weights by user clicks as videos have different intra-weight of relevance. An attention fusion function combines multimodal relevance and recommendations are made accordingly. Vellaichamy et al. [10] proposes a hybrid Collaborative Movie Recommender system that combines Fuzzy C Means clustering with Bat optimization to address the problem of handling huge volumes of data and to enhance clustering accuracy. The system clusters users into different groups and obtains the initial position of clusters using Bat Algorithm to generate relevant movie recommendations. In [11–15] several semantically inclined knowledge centric models in support of the proposed literature have been depicted.

3 Proposed System Architecture

To determine the user's personal preferences, insights must be drawn from the user input, which comprises user queries and personal profile information. The user profile contains the user's historical browsing data. It includes previous searches, previous liked videos, ratings given to videos, videos on the watch list, video channels subscribed by the user, etc. This data allows personalized cognizance of user preferences by analysis of past behavior, allowing the model to learn for a better recommendation. User queries are the present search instances or active inputs from the user regarding his/her preferences. Input data is subjected to preprocessing to provide query terms meaningful for prediction by eliminating unnecessary details. Tokenization, Lemmatization, Stop Word Removal, and Named Entity Recognition are some of the data preprocessing methods available in Python's Natural Language processing library NLTK, which can be used for data preprocessing. The output of this step is tokenized keywords belonging to categorical real-world objects (named entities) and free of redundant data. Preprocessing helps in curating the data for further analysis. The resulting query terms though containing keywords are not enough to describe user preferences, this calls for topic modeling and metadata generation on those query terms, as two steps follow this.

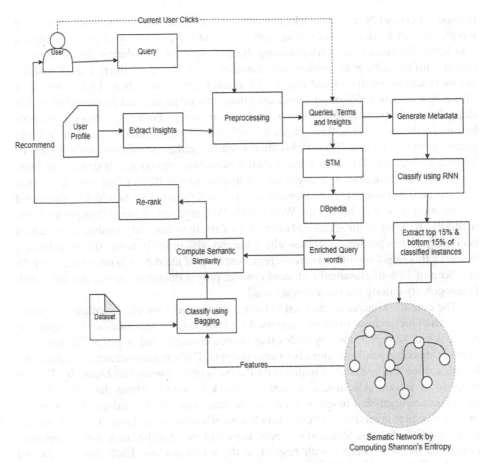

Fig. 1. System Architecture for Proposed SemVidRec

The generated query terms are used to extract metadata by web scraping. A tool is required that analyses data on the semantic web and yields metadata in machine-understandable form. RDF distiller is such a tool. RDF distiller is used to analyze HTML pages annotated by microdata and generate results in RDF specialized formats. This tool examines the content on the HTLM5 pages with the help of microdata. This distiller can be used to search the web for annotated pages containing content, matching the query terms, and obtain metadata in the form of an RDF graph formed by multiple labeled RDF triples. Each triple is made of subject, predicate, and object as its components. The RDF graph obtained as the result of this step helps represent metadata in an unambiguous way. The RDF graph is examined to remove the predicate part of RDF triples, obtaining subjects and objects as separate entities. This is necessary because subjects and objects are to be classified in later steps without considering the connections between them describing the object or data property.

The metadata thus obtained is used exponentially large and there is a need to automatically discover classes from it. For feature identification a deep learning model, RNN is used. Recurrent Neural networks can use internal states to process inputs of varying lengths and make decisions by considering current input along with previous inputs. The RNN discovers classes from among the metadata, by considering the labels associated with the subject and object and classifies them. RNN discovers classes and fits the metadata to the discovered classes. From each discovered class, 15% of the most fitting instances and 15% of the least fitting instances are picked and processed in further steps. This selection scheme is important because, though both categories of instances belong to the same class, there exists a level of diversity in them that can be used to glean insights. The rest 60% of classified instances either are aligned with the top or bottom 15% of the class, contributing least to semantic cognizance. Instances are large in volume and scale, hence it might not be suitable to use 100% of all instances in the recommendation system. The instances are extracted from metadata which is structured information from Web3.0. As the World Wide Web enlarges the metadata grows exponentially. To enhance the speed and efficiency of an algorithm, only a subset of instances is employed. It is possible to choose either the top 30% or the bottom 30% of instances, however this might not offer a diverse representation of the data. Choosing the top 15% and bottom 15% of classified instances ensures proper demarcation and diversity with heterogeneity among the instances selected.

The classified instances are used to create a semantic network. A semantic network is a form of hierarchical ontology map used for knowledge representation. The semantic network can be represented by nodes that encode concepts and edges which represent connections or semantic relations between concepts. The semantic network is cognitively based and the arrangement of nodes is on the basis of a taxonomical hierarchy. For the purpose of feature selection, a semantic network is needed where the selected class instances are organized based on their information measure. The information measure of each instance is realized by computing Shannon's entropy. In the context of instance-based feature selection, Shannon's entropy is used to measure the amount of information contained in each instance with respect to the selected class. Each instance can be thought of as a variable, with its possible values being the class labels. The entropy of an instance is computed by considering the distribution of class labels across the training data that contain that instance. Consider a dataset comprising different instances and their associated class labels. The goal is to identify a subset of instances that are highly informative for a specific class. To achieve this, the entropy value of each instance is calculated with respect to that class using Eq. (1). To compute this entropy value, probabilities that correspond to the frequency of each class label in the set of instances that contain the instance of interest is used.

$$H(X) = -\sum_{i=1}^{n} p(X_i) \log p(X_i) \tag{1}$$

An agent is modeled using AgentSpeak to create a semantic network based on Shannon's entropy. The state of the model is to calculate Shannon's entropy and the behavior is to create links between discovered class instances to form an information tree. This agent is run intra class to get vertical dependencies, later horizontal dependencies between classes are deduced. The agent is parallel processing hence, links are formed faster and the semantic network is formulated quicker. Topic modeling is a statistical model aimed at studying the semantic structure of document collections. This unsupervised probabilistic model helps in identifying latent topics and extracting them from unstructured data. Latent Dirichlet Allocation (LDA) is a commonly used type of topic modeling. However, LDA doesn't allow the inclusion of metadata in modeling topics but only considers the text. Hence Structural Topic Modelling is used to include covariates in influencing the topic prevalence and content in query terms. STM is applied to query terms obtained after preprocessing, yielding relevant topics which might be of interest. These uncovered topics are insights drawn from the user data, useful in determining the interests and needs of the user. A python library version of Structural topic modeling will be used for topic discovery. STM will allow topics, relevant to the query terms to be extracted, however, this will not be enough to enrich the query terms. For the purpose of diversifying the query terms, a knowledge store named DBpedia is used. DBpedia dataset has entities classified as ontology, which are structured information from Wikipedia consisting of real-world facts and data, easily accessible on the web. The Virtuoso infrastructure on which the DBpedia dataset is based allows access to DBpedia RDF data through SPARQL endpoints. Using SPARQL endpoints can enrich the query terms with relevant data.

The enriched query terms are labeled categorical dataset which has to be classified based on the features yielded by the semantic network. The features from the formulated semantic network have to be selected to classify the dataset, for which a bagging classifier is used. In bagging, an ensemble of the outputs from the Random forest classifier and Decision trees is used to improve feature selection. Based on the information measure of each feature, indicated by Shannon's entropy features are selected. Within a decision tree, the features are represented by internal nodes, decision rules are represented by branches, and the classification output is stored in the leaf nodes. So here the decision is based on the level occupied by the feature in the semantic network. The random forest chooses the best prediction from among multiple random decisions previously made. Using features/class selection on each bootstrap will produce the best results. The feature which contains the most useful insights is selected first. The classes and their instances from the Semantic network are classified by the bagging classifier.

To produce suggestions, the enriched query terms are compared to the categorized class labels using measures of semantic similarity. These similarity metrics are obtained through the utilization of two functions, specifically the Modified Twitter Semantic Similarity and the Modified Simpson's Diversity Index. The Twitter Semantic Similarity (TSS) is a semantic similarity model that estimates similarity between words with high precision. The Modified Twitter Semantic Similarity is utilized to investigate the semantic structure of text by analyzing the frequency of word co-occurrence in a document corpus.

The frequency of co-occurrence of words w1 and w2 in tweets, irrespective of the order is calculated according to Eq. 2. $\Phi(w)$ is a measure of the frequency of a word on Twitter based on the velocity of occurrence. Timestamps of tweets are used to calculate the velocity of occurrence. However, in the case of the Semantic video recommendation model, there is no significance of timestamps, so the Modified Twitter Semantic Similarity measure is used which incorporates the deviational Probability distribution of the word in the web corpus. This is a measure distribution of the word across different web pages on the video content platform. TSS is similar in performance to corresponding semantic similarity measures like cosine distance of Latent Similarity and the standard measure of WordNet. The result of Modified TSS is a matrix, where each entry represents the TSS measure between a class instance and the query term. The threshold of 0.75 is used to select the class instances similar to enriched query terms. In conventional and traditional practice, a preferred value of 0.75 is commonly used for various semantic similarity measures like Jaccard similarity and cosine similarity. The TSS measure also falls within this range, yielding values between 0 and 1. The specific value of TSS is influenced by factors such as the model's strength, the number of instances, and the defined epoch set. TSS is computed only once, not iteratively, primarily due to the robustness of the semantic model and the extensive scale of instances involved. Therefore, TSS value of 0.75 is the most appropriate choice for the algorithm

$$TSS(w1, w2) = \left(\frac{\phi(w1, w2)}{max(\phi(w1), \phi(w1))} \right)^{\alpha} \tag{2}$$

To quantitatively measure the diversity of recommendations, taking into account richness and divergence among the class instances selected, Simpson's Diversity Index is used, with certain modifications. The existing index measures the probability of any two randomly chosen identities in a dataset, belonging to the same type. According to Eq. (2) The SDI is the weighted arithmetic mean of propositional abundances. The proportional abundances take values between 0 and 1 hence SDI also ranges between 0 and 1. To choose a diverse set of classified class instances from various sources, certain modifications must be applied to Simpson's Diversity Index as shown in Eq. (3). Instead of using the square of propositional abundance, one-third of the total product of the square of APMI measure, self-information of enriched query terms, and self-information of specific class instances is used. Step deviation is used for standardizing the abundance values and give more weight to rare species, which can improve the sensitivity of this diversity index. Step deviation of 0.25 ensures diversity and variety in each class.

$$\lambda = \sum_{i=1}^{R} (s_i^2 * H(x) * H(y)) \tag{3}$$

TSS provides a measure of closeness between a query term and a set of class instances. Simpson's Diversity Index is used to measure the diversity of recommendations, considering diversity among the class instances selected. The adjustments made to Simpson's Diversity Index allow for the selection of the most varied group of classified class instances from varied sources. By using a combination of TSS and Simpson's Diversity Index, the recommendation system can provide a set of class instances that are both relevant and diverse, thus providing a more comprehensive set of recommendations to the user. The process of reordering the recommendations is carried out by utilizing semantic similarity values computed by any measure of semantic similarities, such as the cosine similarity measure, to enhance the accuracy of the recommendations. This step aids in further refining the relevance of the suggestions.

If the user is dissatisfied with the recommendations, the system can use user clicks as feedback for improvisation. The user clicks are used as further input to the model to improve the relevance and diversity of the recommendations, based on the user prefer-ences. This method allows the algorithm to learn and adapt to the user's changing needs and preferences by generating a new set of recommendations based on the updated feedback. This process can continue until the user is satisfied with the recommenda-tions, indicated by a lack of further clicks. The algorithm can then stop. Incorporating user feedback through clicks can help to improve the effectiveness of recommendation systems and provide a more personalized experience to users. Figure 1 illustrates the proposed System architecture for SemVidRec Video recommendation model.

4   Performance Evaluation and Results

The study utilizes two distinct datasets, the MMTF-14K, a Multifaceted Movie Trailer Dataset for Recommendation and Retrieval [17] and Video Recommendation System Dataset by GlobeIT Solutions, Pune [18]. Both datasets were strategically integrated into a single unit. The datasets were annotated individually using customized focus crawlers with the world wide web as the reference corpora. The annotations were used to categorize the entities in the datasets, and additional categories were added based on these annotations. The entities were then prioritized. These two datasets were combined into a single, larger dataset that was suitable for both movie and video recommenda-tions, including movie trailers. The effectiveness of the SemVidRec model is evaluated by comparing its performance with other established video recommendation models such as VAVR, PVRRC, and CCVR. The results of the comparison are illustrated in Fig. 2. To assess and compare the performance of the system, a fixed set of queries (7156 queries) is used in the experiment. The baseline models and proposed SemVidRec model are implemented in the same environment and compared on the basis of metrics like Precision, Recall, Accuracy, F-measure, False Discovery Rate (FD), and Normal-ized Discounted Cumulative Gain (NDCG). Precision, Recall, Accuracy, and F-measure are measures of the relevance of predictions made by the model. nDCG quantifies the diversity yielded results. It is evident from Fig. 2, that for Visually-Aware Video rec-ommendation (VAVR) with a cold start, the precision is 92.18%, Recall is 94.08%, Accuracy is 93.13%, F-measure is 93.12%, FDR is 0.08 and nDCG is 0.88. For Person-alized Video Recommendation using Rich Contents (PVRRC), the precision is 92.44%, Recall is 95.07%, Accuracy is 93.76%, F-measure is 93.74%, FDR is 0.08 and nDCG is 0.89. For Cross-Curriculum Video Recommendation (CCVR) Algorithm based on a Video-Associated Knowledge Map the precision is 91.03%, Recall is 94.47%, Accuracy is 93.76%, F-measure is 92.72%, FDR is 0.09 and nDCG is 0.95. For SVM+ Fuzzy C-Means Clustering+ Cosine Similarity model the precision is 90.14%, Recall is 92.68%, Accuracy is 91.41%, F-measure is 91.39%, FDR is 0.1 and nDCG is 0.81.

The proposed SemVidRec model yields the highest precision of 95.43%, highest Recall of 97.47%, highest Accuracy of 96.45%, highest F-measure of 96.44%, lowest FDR of 0.05, and nDCG is 0.97. SemVidRec yields the most relevant yet diverse pre-dictions from the dataset and performs better than the aforementioned baseline models. This is because it incorporates both Auxiliary data in the form of Metadata and supple-mentary knowledge in the form of enriched entities. Structural Topic modeling is used to uncover hidden topics relevant yet not easily recognizable in the queries. Then Dbpedia

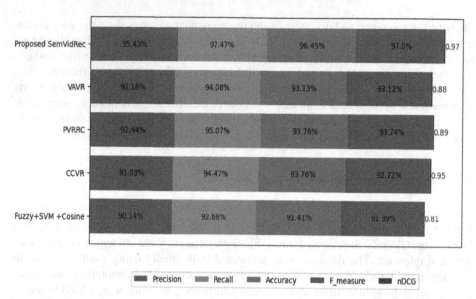

Fig. 2. Performance comparison of Proposed SemVidRec with referential models

knowledge store is used to enrich the query terms hence the nDCG value is very high for the proposed SemVidRec model. The classification of the dataset is achieved by utilizing the features extracted from the semantic network, followed by the implementation of a robust bagging classifier. Bagging ensures a comparative classification model with metadata as features.

The dataset is classified through a bagging classifier, which is an ensemble of decision trees and random forest classifiers. RNN classifies the metadata into categorical and relevant information to be used as features. By computing the Semantic network using Shannon's entropy of the classified metadata, the extraction of highly significant features can be ensured for selection by the bagging classifier. The relevance of prediction made with respect to the user profile and queries is high due to the hybridization of two semantic similarity measures namely Modified Twitter Semantic Similarity and Modified Simpson's diversity index. Both applied with varied threshold and step deviation amount to relevant terms from an ocean of enriched query terms.

VAVR solves the cold start problem, where little annotations are available on a new video, VAVR automatically annotates the videos with visual tags without human intervention. These visual descriptive tags extracted are utilized to generate predictions based on the correlation of visual tags on existing and newly added videos. For extraction of visual features, video is segmented into individual frames based on Color-Histogram distance. As the video is divided into frames, the overall essence of the video is lost and the annotated visual tags are not very relevant in the case of a diverse set of videos. The tags are automatically recognized by machine cognizance but there is a limited amount of information provided to the model. Though the model works in moderate cold start cases, it is highly complex and sometimes yields less relevant predictions based on the density of visual tags discovered. Hence the model has a moderate nDCG value of 0.88 and an accuracy of 93.13%

PVRRC model ensures the personalization from the content in the videos. The model uses collaborative embedded regression to deal with the unavailability of one specific content by integrating a single content feature into collaborative filtering. The model uses rich content both textual and non-textual for the recommendation of videos in both in-matrix and out-of-matrix scenarios. PVRR uses Collaborative Embedded Regression to deal with the unavailability of single specific features. Priority Late Infusion method PRI is used to combine multiple heterogeneous content features both textual and non-textual. The incorporation of CEF and PRI contributes to improving the accuracy hence the model has above average accuracy of 93.76%. The model only considers titles, description, and reviews for textual features and a combination of normalized color histogram and aural tempos as non-textual features. This model fails to incorporate auxiliary information in the form of metadata and supplementary information in the form of enriched query terms, hence the nDCG value is 0.89, while that of SemVidRec is 0.97. This model depends on the knowledge of videos in the dataset which might not always be available and the relevance computation model is not very strong. However, personalization is an

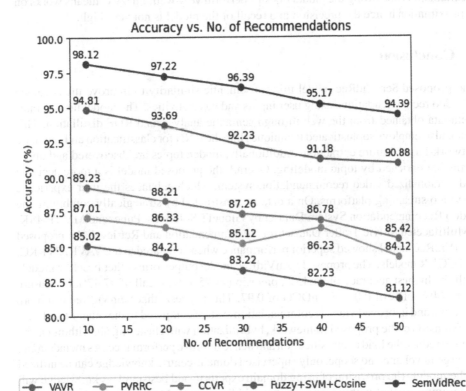

Fig. 3. Accuracy vs Number of Recommendations

added advantage. The relationship between accuracy and a number of recommendations made by SemVidRec and other baseline models is shown in Fig. 3.

The CCVR model aims to help online learners find appropriate learning resources through techniques like the creation of a seed video set, the calculation of correlations between course videos, and the generation of cross-curriculum video subgraphs. By using a video seed set to extract features from student learning platforms and creating a knowledge graph from relevant subgraphs, the CCVR model facilitates efficient resource recommendation for online learning. CCVR model has the highest nDCG value among the baseline models because it includes auxiliary knowledge on existing static knowledge in the form of a knowledge graph. Collaborative filtering is used to filter the generated seed set of videos based on ratings and other feedback. However, not all the videos on the video conglomerate platforms need to be rated. However, due to the inclusion of auxiliary information in the form of a knowledge graph, the model has a high nDCG value is 0.95. The combination of SVM+ Fuzzy C-Means Clustering+ Cosine Similarity model ensures a combination of the binary linear classifier with a strong clustering algorithm and semantic similarity measure. The model lacks auxiliary knowledge and hence has a low nDCG value of 0.81. Since a naive classifier is used and the regulatory mechanism is not strong the model doesn't perform very well. Fuzzy C means works on approximation hence the precision and recall of the model is not very high.

5 Conclusion

The proposed SemVidRec model utilizes semantic similarity to improve the accuracy of video recommendations using user inputs and past activities. The model incorporates metadata obtained from the web through semantic analysis and RDF distillation. This mode also employs sophisticated techniques like the RNN for classification and semantic network-based feature extraction. Additionally, hidden topics are discovered and query terms are enriched by topic modeling. Overall, the proposed model is a more accurate and personalized video recommendation system, which enhances the user experience on video streaming platforms. On a dataset constructed by strategically combining the Video Recommendation System Dataset by GlobeIT Solutions, Pune, and MMTF-14K, a Multifaceted Movie Trailer Dataset for Recommendation and Retrieval, the proposed SemVidRec model showed superior performance when compared to the VAVR, PVRRC, and CCVR models. The proposed SemVidRec model outperforms other baseline models with the highest accuracy of 96.45%, precision of 95.43%, recall of 97.47% F-measure of 96.44%, FDR of 0.05, and nDCG of 0.97. This suggests that SemVidRec is a more effective and improved video recommendation system compared to the other models. As an extension of the proposed framework, hybridization with clustering algorithms can be used to reduce the load of classification and hence improve performance. As metadata can be large in volume and scope, only supervised domain-centric knowledge can be utilized in the model. The provision of using domain experts to add best-in-class knowledge will allow improvement with a human-in-the-middle strategy. Most importantly the overall algorithm can be improved by eliminating the learning algorithms and substituting them with theories like gamification.

References

1. Elahi, M., Hosseini, R., Rimaz, M.H., Moghaddam, F.B., Trattner, C.: Visually-aware video recommendation in the cold start. In: Proceedings of the 31st ACM Conference on Hypertext and Social Media, pp. 225–229 (2020)
2. Du, X., Yin, H., Chen, L., Wang, Y., Yang, Y., Zhou, X.: Personalized video recommendation using rich content from videos. IEEE Trans. Knowl. Data Eng. **32**(3), 492–505 (2018)
3. Zhu, H., et al.: A cross-curriculum video recommendation algorithm based on a video-associated knowledge map. IEEE Access **6**, 57562–57571 (2018)
4. Vimala, S.V., Vivekanandan, K.: A Kullback-Leibler divergence-based fuzzy C-means clustering for enhancing the potential of a movie recommendation system. SN Appl. Sci. **1**, 1–11 (2019)
5. Yan, W., Wang, D., Cao, M., Liu, J.: Deep auto-encoder model with convolutional text networks for video recommendation. IEEE Access **7**, 40333–40346 (2019)
6. Cai, J.J., Tang, J., Chen, Q.G., Hu, Y., Wang, X., Huang, S.J.: Multi-view active learning for video recommendation. In: IJCAI, pp. 2053–2059 (2019)
7. Ma, J., Li, G., Zhong, M., Zhao, X., Zhu, L., Li, X.: LGA: latent genre-aware micro-video recommendation on social media. Multimedia Tools Appl. **77**(3), 2991–3008 (2018)
8. Bhatt, C., Cooper, M., Zhao, J.: SeqSense: Video recommendation using topic sequence mining. In: Schoeffmann, K., et al. (eds.) MMM 2018. LNCS, vol. 10705, pp. 252–263. Springer, Cham (2018). https://doi.org/10.1007/978-3-319-73600-6_22
9. Mei, T., Yang, B., Hua, X.S., Yang, L., Yang, S.Q., Li, S.: Videoreach: an online video recommendation system. In: Proceedings of the 30th Annual International ACM SIGIR Conference on Research and Development in Information Retrieval, pp. 767–768 (2007)
10. Vellaichamy, V., Kalimuthu, V.: Hybrid collaborative movie recommender system using clustering and bat optimization. Int. J. Intell. Eng. Syst **10**(1), 38–47 (2017)
11. Santhanavijayan, A., Naresh Kumar, D., Deepak, Gerard: A semantic-aware strategy for automatic speech recognition incorporating deep learning models. In: Satapathy, S.C., Bhateja, V., Janakiramaiah, B., Chen, Y.-W. (eds.) Intelligent System Design. AISC, vol. 1171, pp. 247–254. Springer, Singapore (2021). https://doi.org/10.1007/978-981-15-5400-1_25
12. Praveen, S.V., Ittamalla, R., Deepak, G.: Analyzing the attitude of Indian citizens towards COVID-19 vaccine–a text analytics study. Diabetes Metab. Syndr. **15**(2), 595–599 (2021)
13. Deepak, G., Gulzar, Z., Leema, A.A.: An intelligent system for modeling and evaluation of domain ontologies for Crystallography as a prospective domain with a focus on their retrieval. Comput. Electr. Eng. **96**, 107604 (2021)
14. Deepak, G., Rooban, S., Santhanavijayan, A.: A knowledge centric hybridized approach for crime classification incorporating deep bi-LSTM neural network. Multimed Tools Appl **80**, 28061–28085 (2021)
15. Kumar, A., Deepak, G., Santhanavijayan, A.: HeTOnto: a novel approach for conceptualization, modeling, visualization, and formalization of domain centric ontologies for heat transfer. In: 2020 IEEE International Conference on Electronics, Computing and Communication Technologies (CONECCT), pp. 1–6. IEEE (2020)
16. Umaa Mageswari, S., Mala, C., Santhanavijayan, A., Deepak, G.: A non-collaborative approach for modeling ontologies for a generic IoT lab architecture. J. Inf. Optim. Sci. **41**(2), 395–402 (2020)
17. Deldjoo, Y., Constantin, M.G., Ionescu, B., Schedl, M., Cremonesi, P.: MMTF-14K: a multifaceted movie trailer feature dataset for recommendation and retrieval. In: Proceedings of the 9th ACM Multimedia Systems Conference, pp. 450–455 (2018)
18. GlobeIT Solutions, Pune, India. Video Recommendation System Dataset (2015). https://www.datahub.io/dataset/video-recommendation-system-dataset

HeartBeatNet: Unleashing the Power of Attention in Cardiology

Gurjot Singh, Anant Mehta[(⊠)], and Vinay Arora

CSED, Thapar Institute of Engineering and Technology, Patiala, Punjab, India
{gsingh_be20,amehta1_be20,vinay.arora}@thapar.edu

Abstract. Cardiovascular disease diagnosis and prompt medical care depend critically on the classification of heart sounds. In recent years, deep learning-based approaches have shown promising results in automating the process of heart sound categorization. This paper proposes a model HeartBeatNet (an attention UNet-based system) for heart sound classification that demonstrates comparatively better performance. The proposed system combines the strengths of attention mechanisms and the UNet architecture to effectively capture relevant features and to make accurate predictions. The system is trained on the PhysioNet/Cinc 2016 dataset consisting of annotated heart sound recordings, which are first converted into Mel Spectrograms before feeding into the UNet based network. The results indicate that the proposed system achieves high accuracy of 95.14%, sensitivity of 90.00%, and specificity of around 96.72% to classify various heart sound abnormalities.

Keywords: Deep Learning · Neural Network · Mel Spectrograms · Phonocardiograms · Feature Extraction

1 Introduction

Heart health is a crucial aspect of overall wellness. The heart circulates nutrient-rich blood, and supplies oxygen. The size of the heart, which is a muscular hollow organ, is comparable to a closed fist [1]. The four chambers of the human heart are the right atrium, right ventricle, left atrium, and left ventricle. Cardiac sounds play a key role in the research of cardiac physiology. Globally, cardiovascular diseases persist to be the primary cause of death and disability. In order to effectively manage and treat these illnesses, prompt and precise diagnosis is absolutely essential. Heart sound analysis, commonly referred to as phonocardiography, has become a non-invasive and economical method for identifying a variety of heart diseases [2]. However, due to the complexity of the data and the presence of numerous physiological and pathological factors that may alter their features, interpreting heart sounds continues to be a difficult task. In recent years, machine learning and deep learning techniques have shown great potential for automating the process of heart sound analysis. These approaches aim to develop algorithms capable of accurately classifying heart sounds into

© The Author(s), under exclusive license to Springer Nature Switzerland AG 2024
R. Muthalagu et al. (Eds.): CINS 2023, CCIS 1978, pp. 14–25, 2024.
https://doi.org/10.1007/978-3-031-48984-6_2

normal and abnormal categories [3]. The semantic segmentation architectures have drawn a lot of attention among the numerous examined and investigated machine learning models because of their success in medical picture segmentation tasks [4]. Mel Spectrograms provide a promising method for automated analysis, utilising the ability of frequency-based features to identify various heart problems and facilitate precise diagnosis [5]. This method has a lot of potential to enhance the diagnosis and treatment of cardiovascular illness by identifying the different patterns in heart sounds. Semantic segmentation of mel spectrograms enables the precise identification and localization of different components within heart sounds, facilitating a detailed analysis and enhancing the understanding of underlying cardiac abnormalities [6]. The U-Net architecture, one of the many deep learning models, has drawn a lot of interest as it can segment medical images [7].

In this research manuscript, the authors have proposed an enhanced heart sound classification framework using an attention-based U-Net model. The U-Net architecture, initially developed for image segmentation, has shown promising results for capturing both local and global features in biomedical signals [8]. By incorporating attention mechanisms into the U-Net model, the authors have aimed to improve the discriminative power of the model and enhanced the focus on relevant regions of the heart sound signals.

2   Related Work

In 2023, Gharehbaghi et al. reported the findings of a study that was carried out on a Parallel Convolutional Neural Network (PCNN) with the intention to identify abnormal cardiac conditions based on the heart sound signals. On the Physionet/Cinc 2016 dataset, the PCNN's accuracy was judged to be 87.20% [9]. In order to merge the characteristics from the chromagram and the textural features from the existing spectrogram, Taneja et al. in 2023, retrieved textural features including linear binary pattern (LBP), adaptive-LBP, and ring-LBP. The accuracy increased with the combination of features taken from both image variations. The experiment produced mean values of 94.87%, 93.11%, and 95.27% for accuracy, precision, and F1-score, respectively [10]. In 2020, Malik et al. proposed effective and highly precise method for heart sound categorization. 2D scalogram images were generated using the wavelet transform method which were further fed to the convolutional neural network-based classifier. The study investigated binary as well as multi-class classification [11].

Ren et al. in 2018 examined the effectiveness of learning deep PCG representations and leveraging pre-trained CNNs from large-scale image data to classify PCG signals. The scalograms were fed into a complete CNN created by transferring learning to a pre-trained network. Findings showed that on the heart sound classification challenge, deep PCG representations generated from a tailored CNN performed best, with a mean accuracy of 56.20% [12].

In 2018, Meintjes et al. used continuous wavelet transform (CWT) scalograms images as an input to CNN to study classification of the basic heart sounds. Support Vector Machine (SVM), CNN, and K-Nearest Neighbor (KNN) were used to distinguish both the initial and subsequent cardiac sounds from the generated images, and their performance was compared. For each PCG signal the magnitude scalogram was obtained, and the CNNs were trained and tested using these. The said research resulted in a specificity of 86.70% and an accuracy of 86.00% [13]. In 2019, Demir *et al.* developed a technique for the diagnosis of heart disorders that used three sequential steps *viz.* spectrogram production, deep feature extraction, and heart sound categorization. In the third stage, an SVM classifier was utilised to extract the deep features from three distinct pre-trained CNNs, *i.e.* AlexNet, VGG16, and VGG19 [14]. Zabihi *et al.* in 2016 devised a method for automatically categorising PCG recordings without segmentation to detect anomalies (normal vs. abnormal) and quality (good vs. bad). The suggested approach attained overall accuracy of 91.50% and 85.90% on the training and unknown testing datasets, respectively [15].

Kay *et al.* in 2017 used a hidden semi-Markov model-based open-source technique to partition the heart sounds. Then, using a new application of the CWT, mel-frequency cepstral coefficients, and some complexity measurements, the time-frequency pattern of an isolated heartbeat was described. Additional elements were also retrieved to describe the behaviour of the heart sounds between beats. The accuracy of this approach on testing data was 85.20% [16]. In 2022, Mashhoor *et al.* provided a Siamese network architecture to distinguish between normal and abnormal signals to find similarity among similar normal and abnormal signals. By implementing this distinction and similarity learning throughout all domains, the challenge of domain-invariant heartbeat classification could be effectively addressed. The multi-domain 2016 Physionet/CinC challenge dataset was utilised by the authors to evaluate the given approach. On the evaluation set the technique achieved an accuracy of 79.10%, a sensitivity of 82.8 percent, and a specificity of 75.30% [17].

In 2016, a variation of the AdaBoost classifier was implemented by Potes *et al.* that received 124 time-frequency features, which were retrieved from the PCG signals. Four frequency ranges of PCG cardiac cycles were utilized to train a second CNN classifier. The final decision rule for categorizing normal and aberrant cardiac sounds was based on an ensemble of classifiers utilizing the AdaBoost and CNN outputs. Sensitivity, specificity, and total score for the classifier ensemble approach were 94.24%, 77.81%, and 86.02%, respectively [18]. In 2021, Huai *et al.* proposed a convolutional neural network-based system for classifying heart sounds. The cardiac sounds were converted into 5-second-long grayscale photographs. An accuracy of 94.80% was attained using the CNN-based technique [19]. In 2017, Rubin *et al.* used deep learning to automate the cardiac inspection process. The authors presented a deep CNN and time-frequency heat map representations-based automatic heart sound categorization system. Given the cost-sensitive nature of misclassification, a modified loss function that directly optimises the trade-off between sensitivity and specificity was used to train the CNN architecture. The technique achieved an accuracy of 83.99% [20].

CNNs are preferred for general image classification, but UNet is superior for jobs involving pixel-by-pixel classification, such as semantic segmentation. Mel-Spectrogram classification does include pixel-by-pixel analysis, therefore using a deep CNN and UNet architecture produces superior outcomes. Hidden Markov Models are better suited for sequential data analysis and aren't commonly used for image classification tasks. Similarly, ANNs are generally not used for image classification tasks as they do not account for spatial information in the image.

3 Methodology

Figure 1 shows an exhaustive understanding of the subject matter by illustrating the steps utilized in the proposed work. A detailed discussion of these steps is given below.

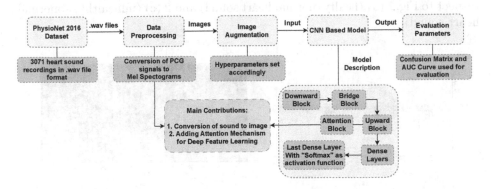

Fig. 1. Sequential steps followed in the proposed work to achieve the classification.

3.1 Dataset Description

For this study, the authors have utilized the PhysioNet 2016 dataset that provides a comprehensive compilation of heart sound recordings obtained from diverse global sources contributed by individuals around the world. This dataset includes recordings from healthy individuals as well as patients with various cardiac conditions. The duration of the .wav recordings ranged from 5 to 120 s. Dataset titled Training-a, Training-b, Training-c, and Training-e have been taken for experimentation in this research work. Collectively, these subsets contained 3071 recordings of the heart sounds [21]. Each individual, whether healthy or pathological, has contributed between one and six different cardiac sound recordings. To assure uniformity, all recordings were resampled and their frequencies were adjusted to 2,000 Hz. Particularly, each recording contained only a single PCG lead, which facilitated the diagnostic process [21].

3.2 Data Pre-processing and Image Augmentation

For the purpose of feature extraction and classification, Mel-spectrograms were generated from the combined PCG signal data. The sample rate, hop length (number of samples between frames), length of the Fast Fourier Transform (FFT), window size, and subplot grid parameters were set properly during the loading phase. The train-to-validation ratio was set to 80:20 and a train-to-test ratio was set to 73:27. To improve the model's robustness, image augmentation techniques were employed [22]. The parameters for these modifications, such as horizontal flip, filling mode, batch size, seed value, sheer range (which defines the angle of slant in the counter-clockwise direction), zoom range (randomly enlargement in or out of the image), and float value (provided as a parameter in the associated augmentation module), were configured appropriately. In addition, the training images were randomized prior to being fed into the neural network. Figure 2 (b) and 2 (d) depict the respective Mel-spectrograms generated with respect to Fig. 2 (a) (healty/normal heart sound) and 2 (c) (unhealthy/abnormal heart sound).

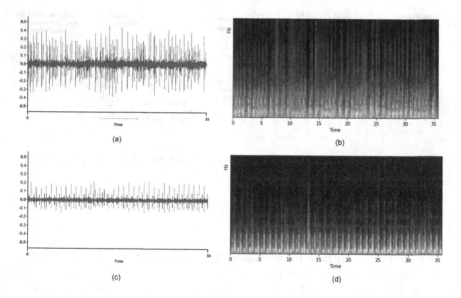

Fig. 2. Phonocardiogram singal (a) and (c) with the respective Mel-Spectrogram (b) and (d).

3.3 Proposed Model

General UNet Architecture. UNet contains three primary elements: the downward path, the bridge, and the upward path [23]. The downward path, is also known as the contracting path, and it is made up of pooling layers and 2-D convolutional layers. It analyzes the input image data and extracts the high-level features, which are then encoded. By establishing a connection between the lower and higher networks, the bridge finishes off the information flow. On the other hand, the upward path, which is also known as the expanding path, is made up of standard and transposed 2-D convolutional layers. It takes the encoded features that were obtained from the downward path and performs an upsampling operation on them [24].

Proposed HeartBeatNet Architecture. In the proposed work, an attention block was incorporated into the UNet architecture's upward block. By emphasizing important features selectively during the upsampling procedure, this enhancement enabled us to further improve the segmentation performance of the UNet model. By incorporating the attention mechanism into the upward block, the authors here have aimed to improve the model's ability to capture and emphasize relevant information while suppressing irrelevant details. With the addition of the attention block, the authors have exploited the spatial dependencies in the maps of features and dynamically weight based on their importance. This attention mechanism improved the accuracy and precision of segmentation map predictions by facilitating the network's concentration on important regions. The proposed HeartBeatNet architecture has enhanced the visualization and identification of significant image pixels by utilizing contextual and location information from both the downward and upward networks.

$$z = Attention(x) \tag{1}$$

$$g = Downwards(x) \tag{2}$$

$$y = Upwards(z, g) \tag{3}$$

In the aforementioned equations, x stands for the input data, z for the attention mechanism depicted in Eq. 1, g for the encoded features depicted in Eq. 2, and y for the decoder's output depicted in Eq. 3.

Figure 3 depicts the layers and attention block used in the proposed Heart-BeatNet Model. The HeartBeatNet has generated the multidimensional pooling feature map as a convolutional layer, which was then flattened into a one-dimensional vector. A dropout layer was added to prevent overfitting. The output of the HeartBeatNet architecture was then sent to a dense artificial neural network. In a dense layer, each neuron is linked to every neuron in the layer above it. The Leaky Rectified Linear Unit (LeakyReLU) activation function has been used in each dense layer. LeakyReLU is different from traditional ReLU as it can handle negative inputs by giving them a small random slope. Two filters and the softmax activation function are used to add the final dense layer, which has been used as the output layer. Softmax figures out the probability distribution for all possible outcomes and puts the results in a vector whose values add up to 1. The class of the test sample is predicted by the probability with the highest value.

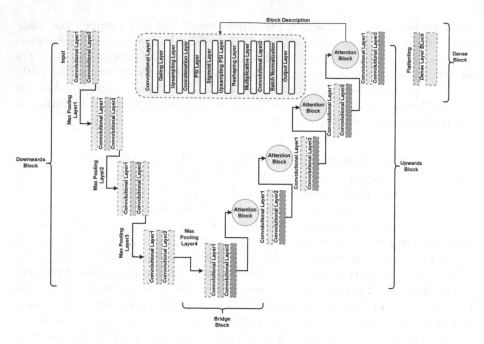

Fig. 3. Sequential Visualization of the proposed HeartBeatNet Architecture based on UNet + Attention based Layers.

4   Results

In a matter of classification, the model's credibility is evaluated based on the confusion matrix, or the predictability of true positives and true negatives. The generated metrics aid in understanding the efficacy and robustness of the model. Figure 4 represents the confusion matrix generated by the proposed model. Using a confusion matrix, the formulas shown in Table 1 can be used to calculate the below mentioned metrics. In this research, metrics such as Accuracy, Precision, Recall, Specificity, and F1 Score have been used. Table 1 lists the performance results of the proposed model on various evaluation metrics. It provides a comprehensive summary of the model's accuracy, precision, recall, specificity, and F1-score.

Table 2 provides a comprehensive comparison between the proposed Heart-BeatNet model and other models that have been implemented thus far in the research field. The accuracy of our proposed HeartBeatNet model surpasses that of all other models, as determined by a thorough analysis.

The relationship between the model's validation accuracy and training accuracy with number of epochs has been depicted in Fig. 5. The graph illustrates a gradual improvement in training as well as validation precision as the number of epochs grows. The consistent increase in training accuracy demonstrates the model's ability to learn and adapt to the used dataset.

Table 1. Results of the proposed HeartBeatNet model

Evaluation Metric	Formula	Proposed Model Values
Accuracy	$\frac{TP+TN}{TP+TN+FP+FN}$	95.14%
Precision	$\frac{TP}{TP+FP}$	88.87%
Recall	$\frac{TP}{TP+FN}$	90.00%
Specificity	$\frac{TN}{TN+FP}$	96.72%
F1 Score	$\frac{2 \cdot Precision * Recall}{Precision+Recall}$	89.44%

Table 2. Comparison of HeartBeatNet with other models

Architecture	Classification Accuracy
Parallel CNN [9]	87.20%
Linear Binary Pattern (LBP) [10]	94.87%
Fine-tuned CNN [12]	56.20%
Continous Wavelet Trans. & CNN [13]	86.00%
Ensemble of ANNs [15]	85.90%
Hidden Semi-Markov model [16]	85.20%
Siamese Network [17]	79.10%
AdaBoost Classifier [18]	86.02%
CNN on Greycale Images [19]	94.80%
Deep CNN on T-F Heat Map [20]	83.99%
Proposed Model (HeartBeatNet)	**95.14%**

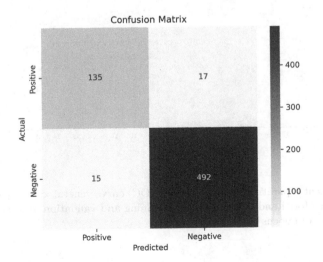

Fig. 4. Confusion Matrix generated by the proposed HeartBeatNet Model, which futher is used to derive the evaluation metrics.

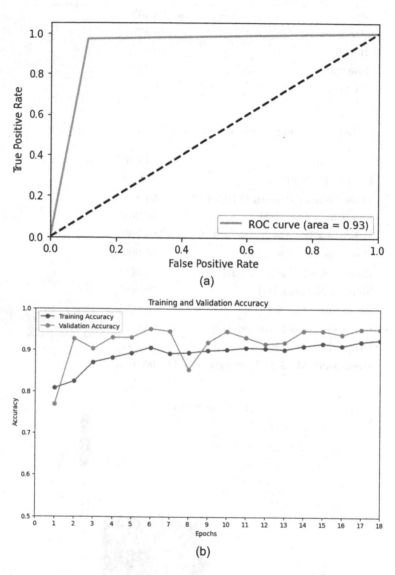

Fig. 5. Recipient Operating Characteristic (ROC) curve generated (a) by the proposed HeartBeatNet Model and variation in the training and validation accuracy depending on the number of epochs (b).

A ROC curve is a graph of probabilities that demonstrates the effectiveness of a classification model across an array of cut-off values. The ROC curve with a notable value of 0.93 is displayed in Fig. 5. This high value indicates that the proposed model has distinguished healthy and murmuring cardiac sounds exceptionally well.

4.1 Experimental Setup

The code for the deep learning architectures was executed on the Nvidia-dgx server, using the A100 GPU. Code was written using Tensorflow (v2.12.0) and Keras. The audio files were converted to spectral-images using Librosa python package. Numpy and Pandas were also essentially used for pre-processing the data. Python (3.0) was used as the primary language to code the deep learning pipeline. For any other task basic inbuilt python libraries were used.

5 Future Scope and Discussion

The application of an attention-based U-Net model in heart sound classification opens up new avenues for early detection and diagnosis of cardiovascular diseases. This strategy may be enhanced and improved with more research and development to give cardiac healthcare practitioners real-time, non-invasive, and affordable tools. This approach might improve clinical practice by aiding in treatment planning, improving diagnostic precision, and promoting proactive management of cardiovascular diseases. The effectiveness of heart sound classification models can be further improved with additional developments in deep learning methodologies, such as the inclusion of multihead-attention processes, transfer learning, and reinforcement learning. This might result in the creation of trustworthy and effective tools for cardiac healthcare specialists, empowering them to make wise choices and give patients individualised treatment. However, one of the significant challenges in heart sound classification is the availability of large and diverse datasets for effective training and validation of these models. High-quality heart sound recordings from a broad spectrum of patients, covering different ages, genders, and ethnicities, are essential. In this context, Generative adversarial networks (GANs) offer a promising solution. GANs are a class of artificial intelligence algorithms that can generate data similar to the input data. For heart sound classification, GANs can be employed to generate artificial heart sound data. This synthetic data can supplement existing datasets, especially in scenarios where real-world data is scarce or imbalanced. By learning the underlying distribution of heart sounds, GANs can produce realistic variations of heart sounds. This not only addresses the data imbalance issue but also potentially enhances the classification model's accuracy and robustness by providing a richer training dataset. In conclusion, heart sound classification using deep learning and attention based techniques has a huge scope for further analysis. This technology has the potential to revolutionise the area of cardiology and improve patient outcomes as well as healthcare delivery with additional study, development, and collaboration.

References

1. Khairy, P., Poirier, N., Mercier, L.-A.: Univentricular heart. Circulation **115**(6), 800–812 (2007)
2. Nabel, E.G.: Cardiovascular disease. N. Engl. J. Med. **349**(1), 60–72 (2003)
3. Wang, J., et al.: Detecting cardiovascular disease from mammograms with deep learning. IEEE Trans. Med. Imaging **36**(5), 1172–1181 (2017)
4. Guo, Y., Liu, Y., Georgiou, T., Lew, M.S.: A review of semantic segmentation using deep neural networks. Int. J. Multimedia Inf. Retrieval **7**, 87–93 (2018)
5. Nguyen, M.T., Lin, W.W., Huang, J.H.: Heart sound classification using deep learning techniques based on log-mel spectrogram. Circ. Syst. Sig. Process. **42**(1), 344–360 (2023)
6. Zhou, Z., Rahman Siddiquee, M.M., Tajbakhsh, N., Liang, J.: UNet++: a nested U-Net architecture for medical image segmentation. In: Stoyanov, D., et al. (eds.) DLMIA/ML-CDS -2018. LNCS, vol. 11045, pp. 3–11. Springer, Cham (2018). https://doi.org/10.1007/978-3-030-00889-5_1
7. Nannan, Yu., He, Yu., Li, H., Ma, N., Chunai, H., Wang, J.: A robust deep learning segmentation method for hematoma volumetric detection in intracerebral hemorrhage. Stroke **53**(1), 167–176 (2022)
8. Ronneberger, O., Fischer, P., Brox, T.: U-Net: convolutional networks for biomedical image segmentation. In: Navab, N., Hornegger, J., Wells, W.M., Frangi, A.F. (eds.) MICCAI 2015. LNCS, vol. 9351, pp. 234–241. Springer, Cham (2015). https://doi.org/10.1007/978-3-319-24574-4_28
9. Gharehbaghi, A., Partovi, E., Babic, A.: Parralel recurrent convolutional neural network for abnormal heart sound classification. CARING IS SHARING-EXPLOITING THE VALUE IN DATA FOR HEALTH AND INNOVATION, pp. 526 (2023)
10. Taneja, K., Arora, V., Verma, K.: Classifying the heart sound signals using textural-based features for an efficient decision support system. Expert Syst. **40**(6), e13246 (2023)
11. Malik, A.E.F., Barin, S., Emin Yüksel, M.: Accurate classification of heart sound signals for cardiovascular disease diagnosis by wavelet analysis and convolutional neural network: preliminary results. In: 2020 28th Signal Processing and Communications Applications Conference (SIU), pp. 1–4. IEEE (2020)
12. Ren, Z., Cummins, N., Pandit, V., Han, J., Qian, K., Schuller, B.: Learning image-based representations for heart sound classification. In: Proceedings of the 2018 International Conference on Digital Health, pp. 143–147 (2018)
13. Meintjes, A., Lowe, A., Legget, M.: Fundamental heart sound classification using the continuous wavelet transform and convolutional neural networks. In: 2018 40th Annual International Conference of the IEEE Engineering in Medicine and Biology Society (EMBC), pp. 409–412. IEEE (2018)
14. Demir, F., Şengür, A., Bajaj, V., Polat, K.: Towards the classification of heart sounds based on convolutional deep neural network. Health Inf. Sci. Syst. **7**, 1–9 (2019)
15. Zabihi, M., Rad, A.B., Kiranyaz, S., Gabbouj, M., Katsaggelos, A.K.: Heart sound anomaly and quality detection using ensemble of neural networks without segmentation. In: 2016 Computing in Cardiology Conference (CinC), pp. 613–616. IEEE (2016)
16. Kay, E., Agarwal, A.: DropConnected neural networks trained on time-frequency and inter-beat features for classifying heart sounds. Physiol. Meas. **38**(8), 1645 (2017)

17. Mashhoor, R.Y., Ayatollahi, A.: HeartSiam: A domain invariant model for heart sound classification. arXiv preprint arXiv:2210.16394 (2022)

18. Potes, C., Parvaneh, S., Rahman, A., Conroy, B.: Ensemble of feature-based and deep learning-based classifiers for detection of abnormal heart sounds. In: 2016 Computing in Cardiology Conference (CinC), pp. 621–624. IEEE (2016)

19. Huai, X., Kitada, S., Choi, D., Siriaraya, P., Kuwahara, N., Ashihara, T.: Heart sound recognition technology based on convolutional neural network. Inform. Health Soc. Care **46**(3), 320–332 (2021)

20. Rubin, J., Abreu, R., Ganguli, A., Nelaturi, S., Matei, I., Sricharan, K.: Recognizing abnormal heart sounds using deep learning. arXiv preprint arXiv:1707.04642 (2017)

21. Reyna, M.A., et al.: Heart murmur detection from phonocardiogram recordings: the George B. moody physionet challenge 2022. In: 2022 Computing in Cardiology (CinC), vol. 498, pp. 1–4. IEEE (2022)

22. Van Dyk, D.A., Meng, X.-L.: The art of data augmentation. J. Comput. Graph. Stat. **10**(1), 1–50 (2001)

23. Moreland, K., Angel, E.: The FFT on a GPU. In: Proceedings of the ACM SIGGRAPH/EUROGRAPHICS Conference on Graphics Hardware, pp. 112–119 (2003)

24. Sivagami, S., Chitra, P., Kailash, G.S.R., Muralidharan, S.R.: UNet architecture based dental panoramic image segmentation. In: 2020 International Conference on Wireless Communications Signal Processing and Networking (WiSPNET), pp. 187–191. IEEE (2020)

EEG-Based Identification of Schizophrenia Using Deep Learning Techniques

B. Shameedha Begum[1]([⊠]), Md Faruk Hossain[1], Jobin Jose[2], and Bhukya Krishnapriya[3]

[1] National Institute of Technology, Tiruchirappalli, Tamil Nadu 620015, India
shameedha@nitt.edu
[2] Indian Institute of Information Technology, Kottayam, Kerala 686635, India
[3] Indian Institute of Information Technology, Design and Manufacturing, Kancheepuram, Tamil Nadu, India

Abstract. A detection system for the diagnosis of Schizophrenia using machine learning and deep learning techniques are used in this study. Schizophrenia is a brain disorder which can be identified by various symptoms. Most common symptoms of Schizophrenia are speech disorder, laughing without any reason, crying without any reason, poor memory, lack of motivation etc. EEG signals are collected from human brains by placing electrodes (metal discs) on the scalp using a device named Electroencephalogram. It measures electrical activity of the brain, and the data is represented in the form of EEG signals. EEG signals are mainly used to study various diseases of the human brain. EEG signals of 14 healthy persons and 14 Schizophrenia patients are used. One machine learning classification algorithm, i.e. logistic regression and two deep learning models, i.e. convolutional neural network (CNN), and combination of multiple layers of convolutional neural networks and gated recurrent unit (GRU) are used to analyze the signals. Manual features are extracted from EEG signals and then feed into logistic regression to classify the signals. Extraction of Mel Frequency Cepstral Coefficient (MFCC) feature is done. Deep learning models are used to classify the EEG signals.

Keywords: Schizophrenia · EEG · MFCC · Features extraction · Logistic Regression · CNN · GRU

1 Introduction

1.1 Background

Diseases can be diagnosed by using biological markers, and doing laboratory tests, or by using imaging techniques. However, based on the interviews with patients and using the behavior of a patient, disease can be diagnosed. Schizophrenia (SZ) is a mental disorder which can be diagnosed in people of any age. It stops normal thinking, behavioral characteristics, and speech of a person. Symptoms of Schizophrenia include speech disorganization, hallucinations, and deterioration of functional work etc. Environmental dangers like low birth weight, premature birth, exposure to a virus at early age of life, and migrant status hampers the development of the brain, which can lead to Schizophrenia. However, since SZ depends on genetic factors, individuals having the genes zinc finger 804 A and Neurogranin have a greater chance of developing it. Life quality needs to be compromised when a patient is suffering from Schizophrenia, because most of the Schizophrenia patients are not able to live a normal life, twenty to forty percent of the patients try to attempt suicide. Therefore, timely diagnosis of Schizophrenia is very important. Schizophrenia Patients desire a good treatment for the recovery. Brain waves have various bands of signals. EEG signals are categorized into five basic groups: delta band (1–4 Hz), theta band (4–8 Hz), alpha band (8–12 Hz), beta band (12–30 Hz), gamma band (above 30 Hz). When the frequency of EEG signals are too low, it means the person is in deep sleep. And, when the frequency of EEG signals are too high, it means the person is doing multi-tasking. The idea is to identify Schizophrenia patients using EEG (Electroencephalogram) signals. Electroencephalography (EEG) signals are neural activities and generally the integrals of potentials which draw out from the brain with different frequencies. EEG is one of the main diagnostic tests for Schizophrenia. Also, to detect other brain disorders, EEG plays an important role.

1.2 Motivation

One of the popular neuroimaging technique named multi-modal imaging which is used currently to detect Schizophrenia. These processes are MRI scan, CT scan, functional resonance magnetic imaging, and emission positron tomography. But the above methods are very costly to setup and doing the tests. A combination of all these methods can be used when only one method alone not able to detect the brain disease. But, using a combination of all these imaging devices are very costly, and due to motion artifacts, the images coming from the two different devices may not be of a good quality. Therefore, EEG signals can be used to detect Schizophrenia which is more cost-effective. Electroencephalograms (EEG) signals represent electrical activity of the brain collected by placing electrodes on scalp of humans. The method of detecting Schizophrenia using EEG signals are more cost-effective compared to other proposed methods to detect Schizophrenia. In comparison to neuroimaging techniques, when EEG signals are used to detect Schizophrenia, then there is no need for any costly machines like MRI scan and CT scan which are generally used in neuroimaging techniques. Instead of using those costly machines, EEG signals can be used to detect Schizophrenia patients. To collect EEG Signals, only EEG (Electroencephalogram) machine is needed whose cost is much less compared to MRI and CT Scan machines. The models used in this paper gives a good accuracy which can be implemented in real life to detect Schizophrenia patients.

2 Literature Review

D. Ahmedt-Aristizabal et al. [1] the prospective identification of childrens who have more chances of developing schizophrenia is an important thing to do early inventions of Schizophrenia. Classification algorithms like KNN and SVM are compared with LSTM to see the improvement in the accuracy using LSTM model to identify schizophrenia within childrens.

S. Roy et al. [2] it used recently released "TUH Abnormal EEG Corpus" dataset for finding out the performance of the algorithms used. It Focused on proper identification of EEG signals. Various differences can be observed between a normal EEG and abnormal EEG. It pre-processed the EEG dataset and applied mean, median and standard deviation to classify the EEG dataset and use GRU model.

T. Wang et al. [3] it focuses on the initial stage in interpreting EEG signals to check if the activity of brain is normal or abnormal. To solve this task, it proposed a RNN architecture known as ChronoNet which focused on the area of image classification and depending of the features extracted from the images it predict the results.

M. Tanveer et al. [4] based on analyzing the schizophrenia patients based on deep belief networks (DBNs). The proposed method is evaluated in the biomedical research center using the database. The region of ventricle from the normal and Schizophrenia images is divided into segments using a method named as multiplicative intrinsic component optimization.

R. L. Miller et al. [5] dataset consist of MRI images (series of X-ray images) to detect Schizophrenia. Dataset consist of a large MRI images having ($x = 631$, including 246 schizophrenia patients from six imaging resources), and a functional connectivity of features was developed using deep discriminant autoencoder network to make difference between schizophrenic patients from healthy persons.

Analysing EEG data is a cost-effective way to detect Schizophrenia. In CT and MRI scan, a series of brain images are captured. Then features are extracted from those images to detect Schizophrenia, which is less effective in detecting abnormality compared to extracting features from EEG dataset. Analysing EEG signals will give more accurate results compared to images, because EEG signals are collected directly from human brain.

3 Implementation

3.1 Proposed Solution

One machine learning algorithm and two deep learning models are used for detecting Schizophrenia. EEG signals are non-linear in behavior, so, the extraction of the features is done to make difference between the EEG signals of healthy persons and Schizophrenia patients. Since there are only two target classes, i.e; when a person is suffering from Schizophrenia (YES) and another target class is person is healthy (NO), so Logistic Regression is used which is a binary classification machine learning algorithm. Here extraction of mathematical features is done manually and feeds the features in the Logistic Regression. When the frequency of EEG signals are in between 25 Hz to 40 Hz, it means that time person is doing multiple thinking. So, bandpass frequency filter is applied to the EEG signal, and only the signals lies in between 25 Hz to 40 Hz will be considered to train into the models. Due to applying bandpass frequency filter, it is giving more accurate results compared to many previous proposed solutions. MFCC feature extraction is also done before training the models. In MFCC feature extraction process, time-domain EEG signals are converted to frequency-domain EEG signals using Fast Fourier Transform. Deep learning models are also used, i.e. CNN and combination of multiple layers of CNN and GRU. In deep learning, the extraction of features and classification methods both happens by itself. CNN is one of the most powerful technique in deep learning that is attracted by the researchers for the identification of EEG signals which are abnormal and to study those signals to detect disorders like seizure, mental depression and Schizophrenia.

3.2 Block Diagram

First, the raw EEG data is taken. After pre-processing (applying bandpass frequency filter), a pre-processed dataset is generated. From the pre-processed dataset MFCC features and manual features are extracted, and applied one machine learning algorithm and two deep learning models (Fig. 1).

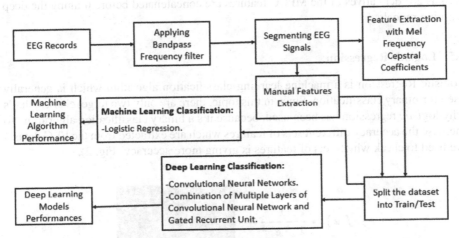

Fig. 1. Block diagram of proposed method

3.3 Data Pre-processing

In the dataset, EEG signals of 14 patients suffering from Schizophrenia, number of male patients were seven and number of female patients were seven, with an average age of 27 and 28 years. And EEG signals of 14 healthy persons were collected of the same age group. So, the dataset contains the EEG signals of 28 people, out of which 14 are Schizophrenia patients and 14 are healthy persons. For each person the data was collected for fifteen minutes at a sampling frequency of 250 Hz. The dataset contains the raw EEG data. At first, preprocessing is done on the EEG data to get the preprocessed dataset. The electrodes (channels) which were used to collect the signals are: Fp2, Fp1, F3, F4, F7, Fz, F8, C3, T3, Cz, T4, C4, T5, T6, Pz, P3, P4, O2, O1.

Since the signals are taken for fifteen minutes, the signals are then divided into segments or epochs. Each segment is of length 25 s, so, the number of samples become 6250. Bandpass frequency filter is applied on the dataset to remove the artifacts and to consider only those signals which are between 25 Hz and 40 Hz.

3.4 Feature Extraction

Features extracted from EEG dataset are: mean, standard deviation, peak-to-peak, variance, minimum value, maximum value, index of minimum value, index of maximum value, kurtosis, skewness and root-mean-square. Features are concatenated and feeded into Logistic Regression. MFCC features are spectrum of frequencies of the signals. In MFCC feature extraction, time-domain EEG signals are converted to frequency-domain EEG signals. The total number of MFCC coefficients used are 13. And, 1st derivatives as well as 2nd derivatives of the MFCC features are concatenated before training the deep learning models.

3.5 Logistic Regression

Logistic Regression is a machine learning classification algorithm which is generally used for binary classification. Since in this topic, there are only two target classes, that's why logistic regression has been used, because it's a binary classification algorithm. To increase the accuracy different sets of features which are extracted from the EEG dataset are used to check which sets of features is giving more accuracy (Fig. 2).

$$f(x) = \frac{1}{1 + e^{-x}}$$

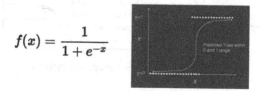

Fig. 2. Logistic Regression function

Logistic Regression use a Logistic/Sigmoid Function to predict the output as 0 or 1, i.e. where, x is the independent variable through which input features are feeded and f(x) is a dependent variable which is the output and it's value always lies in between 0 and 1. A threshold value of 0.5 is taken. So, if the value of f(x) is less than equal to 0.5 then it is considered as 0 (means person is healthy). And if the value of f(x) is more than 0.5 then it is considered as 1 (means person have Schizophrenia).

3.6 Convolutional Neural Networks

Layers in Convolutional Neural Network:

a. Filter (Convolution Layer): In CNN, there are various filters and the values for each filter in the convolutional layer is obtained by training on a particular training set. After the training, a unique set of filter values will be generated that are used to detect specific features in the dataset.

b. Pooling Layer: This layer is for reducing the feature map dimensions. Thus, the parameter's count decreases using this layer. So, it reduces computational work in the network. There are two types of pooling. It finds out all the features present in feature map which is produced by the previous layer.

c. Fully Connected Layer: This layer is a neural network. Each of the neurons uses a linear transformation and feeded in the input vector via a matrix. So, layer-to-layer all possible connections are present, means all vectors of input have effect on every vector of output (Fig. 3).

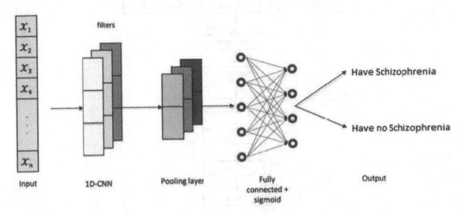

Fig. 3. Layers of Convolutional Neural Networks

The dataset is split into training and test data, then the CNN model training is done. Performance is measured in terms of accuracy.

3.7 Combination of Multiple Layers of CNN and GRU Model

Multiple layers of Convolutional Neural Networks and Gated Recurrent Unit is combined together to build the deep learning model. Output of CNN blocks is feeded as input to the GRU blocks. This deep learning model gives an excellent accuracy.

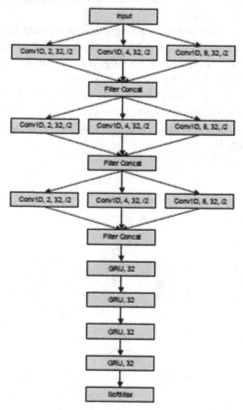

Fig. 4. Combination of multiple layers of CNN and GRU model

The first 3 layers of the model contain CNN blocks. Within each layer 3 CNN blocks are present. After each layer, output of all CNN blocks are concatenated before passing it to the next layer. Total 4 GRU blocks are used in the model with 1 block per layer.

4 Results

Applying the bandpass frequency filter is very much necessary to remove artifacts and external disturbances. Raw EEG signals are shown here, (Collected through 19 channels) before applying the bandpass frequency filter and how they looks after applying the bandpass frequency filter in Figs. 4 and 5.

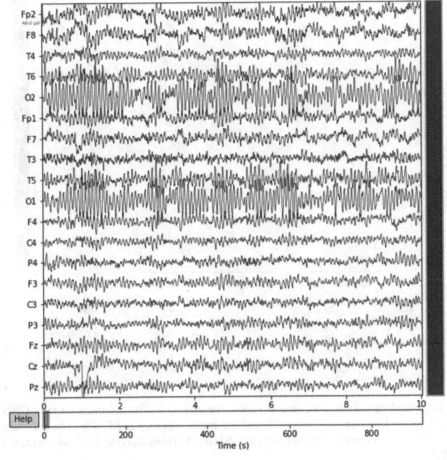

Fig. 5. EEG signals before applying bandpass frequency filter

Here, x-axis represents time and y-axis represents 13 different MFCC coefficients. In the right side, it is the mapping between colours and different numerical values. At each point of the spectrum, it is a value given for MFFC index at a certain point of time (Figs. 6, 7, 8 and 9).

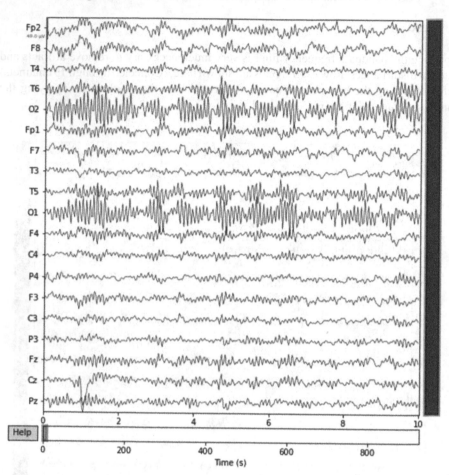

Fig. 6. EEG signals after applying bandpass frequency filter

Logistic Regression gives an accuracy score of 77.3%. The accuracy score of CNN model lies in between 78% and 86% and the mean accuracy is 82.64%. The accuracy score of combination of multiple layers of CNN and GRU model is 96.67% which shown in Table 1

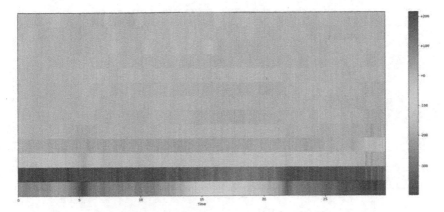

Fig. 7. Spectrum of MFCC features

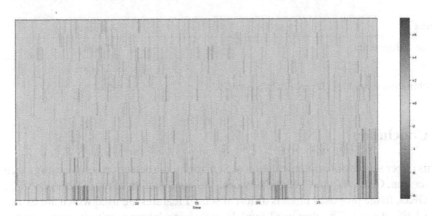

Fig. 8. 1st derivatives of MFCC features

From the above results, it can be clearly seen that combination of multiple layers of CNN and GRU model has outperformedthe Logistic Regression as well as CNN model, because of its multiple layers in the model configuration.

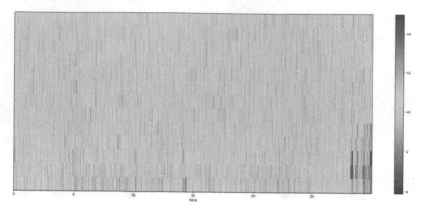

Fig. 9. 2ⁿᵈ derivatives of MFCC features

Table 1. Performance comparison between machine learning & deep learning models

Model	Accuracy
Logistic Regression	77.30%
CNN	82.64%
Combination of multiple layers of CNN and GRU	96.67%

5 Conclusion

In this paper successful implementation of Schizophrenia detection is done using Logistic Regression, Convolutional Neural Network, and combination of multiple layers of CNN and GRU. In the proposed models different sets of features are used which are extracted from EEG dataset. The proposed models are giving better results in many scenarios compared to other models. The proposed models predict good accuracy compared to other Schizophrenia detection techniques.

References

1. Ahmedt-Aristizabal, D., et al.: Identification of children at risk of schizophrenia via deep learning and EEG responses. IEEE J. Biomed. Health Inform. **25**(1), 69–76 (2021). https://doi.org/10.1109/JBHI.2020.2984238
2. Roy, S., Kiral-Kornek, I., Harrer, S.: Deep learning enabled automatic abnormal EEG identification. In: 2018 40th Annual International Conference of the IEEE Engineering in Medicine and Biology Society (EMBC), pp. 2756–2759 (2018). https://doi.org/10.1109/EMBC.2018.8512756
3. Wang, T., Bezerianos, A., Cichocki, A., Li, J.: Multikernel capsule network for schizophrenia identification. IEEE Trans. Cybern. **52**(6), 4741–4750 (2022). https://doi.org/10.1109/TCYB.2020.3035282

4. Tanveer, M., Jatin Jangir, M.A., Ganaie, I.B., Tabish, M., Chhabra, N.: Diagnosis of Schizophrenia: a comprehensive evaluation. IEEE J. Biomed. Health Inform. **27**(3), 1185–1192 (2023). https://doi.org/10.1109/JBHI.2022.3168357

5. Miller, R.L., Yaesoubi, M., Calhoun, V.D.: Cross-frequency rs-fMRI network connectivity patterns manifest differently for schizophrenia patients and healthy controls. IEEE Signal Process. Lett. **23**(8), 1076–1080 (2016). https://doi.org/10.1109/LSP.2016.2585182

6. Huang, Y.-J., et al.: Assessing schizophrenia patients through linguistic and acoustic features using deep learning techniques. IEEE Trans. Neural Syst. Rehabil. Eng. **30**, 947–956 (2022). https://doi.org/10.1109/TNSRE.2022.3163777

7. Arribas, J.I., Calhoun, V.D., Adali, T.: Automatic bayesian classification of healthy controls, bipolar disorder, and schizophrenia using intrinsic connectivity maps from fMRI data. IEEE Trans. Biomed. Eng. **57**(12), 2850–2860 (2010). https://doi.org/10.1109/TBME.2010.2080679

8. Liu, A., Chen, X., Wang, Z.J., Xu, Q., Appel-Cresswell, S., McKeown, M.J.: A genetically informed, group fMRI connectivity modeling approach: application to schizophrenia. IEEE Trans. Biomed. Eng. **61**(3), 946–956 (2014)

9. Singh, K., Singh, S., Malhotra, J.: Spectral features based convolutional neural network for accurate and prompt identification of schizophrenic patients. Proc. Inst. Mech. Eng. [H] **235**(2), 167–184 (2021). https://doi.org/10.1177/0954411920966937

10. Xu, T., Stephane, M., Parhi, K.K.: Abnormal neural oscillations in schizophrenia assessed by spectral power ratio of MEG during word processing. IEEE Trans. Neural Syst. Rehabil. Eng. **24**(11), 1148–1158 (2016). https://doi.org/10.1109/TNSRE.2016.2551700

MLSM: A Metadata Driven Learning Infused Semantics Oriented Model for Web Image Recommendation via Tags

Rishi Rakesh Shrivastava[1] and Gerard Deepak[2(✉)]

[1] Birla Institute of Technology and Science, Pilani, India
[2] Department of Computer Science and Engineering, Manipal Institute of Technology Bengaluru, Manipal Academy of Higher Education, Manipal, India
gerard.deepak.christuni@gmail.com

Abstract. The model proposes image recommendation method by generating tags which is the state of the art in the evolving web 3.0. Proposed model works on the principle of enrichment of queries through topic Modelling and standard knowledge repositories. Data set driven topic synthesis and metadata synthesis by classifying it using Bi-LSTM classifier is the basis for the model. Strong semantic similarity computation measures such as Piyanka index Lance and William index and adaptive pointwise mutual index measures are integrated into the model. An intermediate semantic network is formalized, and Optimization is achieved using the harmonic search algorithm. Proposed MLSM is best among the baseline models with Precision of 94.09% recall of 96.91%.

1 Introduction

In the digital age, we absorb information beyond text. We live in a multimodal world where digital items attract and communicate. Multimedia material augments text. These components engage our readers and offer an appealing alternative to website content. Thus, multimedia material promotes and disseminates knowledge effectively. With the number of people who use the internet rising, we need new ways to spread information that have the most effect. One of these ways is to suggest images by putting tags on them. By giving pictures the right tags or information, we can make them easier to find and reach. Image recommendations based on tags have a number of benefits. First, it lets us use the visual draw of pictures to quickly get people's attention. Images are a powerful way to communicate because they can show complicated ideas and feelings better than words alone. Tagging also makes it possible to group and organize images, which makes it easier for users to look for and find images that match their hobbies. Also, since there is a lot of user-generated content on the web, there needs to be a good way to sort and select pictures. As the amount of multimedia material keeps growing at an exponential rate, it gets harder and harder to directly select and suggest pictures. Automated picture selection systems can help with this by looking at the features of an image, the user's tastes, and the environment to make personalized suggestions. Web 2.0 is quite dynamic and comparably less structured as compared to web 3.0 and so web 2.0 based schemes

R. Muthalagu et al. (Eds.): CINS 2023, CCIS 1978, pp. 38–46, 2024.
https://doi.org/10.1007/978-3-031-48984-6_4

cannot be compared to web3.0 based schemes as the penetrative power of the algorithms can be increased by incorporating knowledge into it.

Contribution: Novel contribution of the proposed Framework is as follows. Growing query by preprocessing it, and application of topic modelling and harnessing entities from Standard knowledge stores, are detailed, enumerated and acclaimed paradigm for improving the query density of the knowledge and contextualizing it. The generated metadata is further transformed as permeable by strong deep learning classifier and optimization is achieved using the harmony search algorithm with strong relevance computation measures.

Organization: the upcoming part of the paper is structured as follows. Section 2 reports the relevant works carried out for image recommendation through tagging. Section 3 contains the proposed system architecture. Section 4 shows the implementation and results of the MLSM model. Section 5 depicts the conclusion drawn from the system, followed by references in Sect. 6.

2 Related Works

Shimizu et al. [1], develops "fashion intelligence" and a visual-semantic embedding system that automatically learns and interprets fashion-related information to answer users' questions. The technique integrates abstract tag information with clothing photos in a common projective environment. Hyvönen et al. [2], proposes a scenario where a user encounters a complex image repository with content that is partially unfamiliar. It highlights the potential benefits of employing ontologies to assist the user in articulating their information requirements, constructing queries, and obtaining relevant answers. To demonstrate the concept's viability, a demonstrational photo exhibition was developed using the Helsinki University Museum's promotion image database and leveraging semantic web technologies. Gao et al. [3], presents a semi-automatic image tagging method assigns each image a category label first. The suggested approach automatically tags images using a massive quantity of online data. In which Sparse coding selects semantically related pictures for effective tag propagation and local and global ranking aggregation approaches increase its resilience to noisy tags. Rawat et al. [4], present a deep neural network that, given a picture and some other contextual information, can make predictions about numerous tags that best describe it. The suggested model is trained in an end-to-end fashion to solve a multi-label classification issue. The model is evaluated using the YFCC100M dataset, which contains 1,965,232 photos and was donated by the Yahoo-Flickr Grand Challenge's event organisers. Rawat et al. [5], proposes a label transfer system to automatically recommend tags for each image using category information. Sparse coding-based spatial pyramid matching and deep convolutional neural networks describe pictures. Metric learning combines these properties to improve picture representation. Hwangbo et al. [6], have proposed a strategic recommendation framework for fashion retail e-commerce which encompasses a collaborative filtering and reflecting on domain characteristics that hybridizes product sale through data which is weighted in order to select the preference's of customers. A decay function is encompassed into the model to track the changes of the products and the preferences over time. De Divitiis et al. [7] have proposed a disentangling feature based framework for fashion recommendation. So the frameworks uses the memory augmented neural networks which helps in recommending variety of garment outfit using given fashion term by considering several co-occurrence of attributes which yields disentangled representation of fashion items stored in the memory model. W. Eom et al. [8] proposes a novel approach for personalized image tag recommendations, leveraging the contextual information of users' favourite images. More specifically, for the task of suggesting tags for recently uploaded images, the method harnesses the tag annotations associated with the favourite images of the uploading user, amalgamating statistical tag data with visual similarity measures. By Lee et al. [9] image tag recommendation problem is addressed through a Maximum a Posteriori (MAP) formulation, incorporating the use of a visual folksonomy. This folksonomy, akin to a collaboratively constructed metadata repository for social categorization, offers significant advantages based on experimental findings. The model uses an extensive and unrestricted concept vocabulary, ultimately resulting in the recommendation of a larger number of accurate tags, thereby enhancing the precision of tag suggestions.

3 Architecture

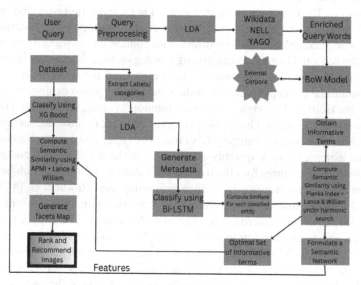

Fig. 1. Proposed System Architecture

The architecture for the proposed knowledge-centric scheme for web image recommendation encompassing hybrid machine intelligence and auxiliary knowledge is depicted in Fig. 1. The user query servers as the input to the framework which is query driven. Although being a knowledge centric framework, the query acts as the first line of input into the framework. User query is subjected to preprocessing which involves tokenization lemmatization stop word removal and name entity recognition, also the word sense disambiguation, is applied for preprocessing the queries. Individual query words are obtained, they are then subjected to topic modelling using LDA(Latent Dirichlet Allocation) which is first applied to a external corpus using the World Wide Web, in order to discover relevant and yet hidden topics for the query words and anchor knowledge to the query words so that it becomes more informative. However, the topics discovered through LDA are quite shallow there is a need for further enrichment so they are further subjected to Wiki Data API and NELL [6] and YAGO [7] knowledge stores which are queried through SPARQL query mechanism to obtain relevant knowledge from the knowledge stores which are then put into enriched query words set. Enriched query words are subjected to bag of words model with external corpora which is crawled from the world wide web considering the entities in the data set which are then used for further recommendation. The data set is a image data set which has labels which is also subjected to LDA modelling. Which are then used for generation of meta data. It is done using the Omeka auto generator which is a extrinsic auto-generator and performs social-tagging. The meta data generated is quiet large so it has to be classified using strong deep learning classifiers. Here we have used Bi-LSTM which is a deep learning model based on LSTM. Two LSTM layers process sequential data in both directions. Capturing previous and future interdependence. The forward LSTM analyses the input sequence left-to-right, whereas the backward LSTM reverses it. Both LSTMs' outputs are merged to reflect the input sequence's past and future. Bi-LSTM can capture complicated relationships and enhance sequence-related tasks as depicted in Fig. 2.

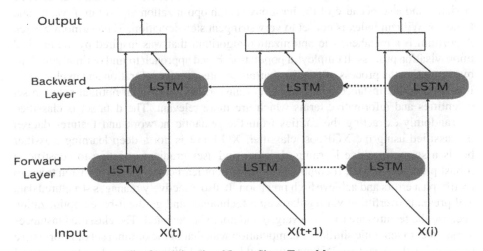

Fig. 2. Bidirectional Long Short-Term Memory.

As it works on the mechanism of auto handcrafted feature selection so for each classified entity under each class of the meta data SimRank is computed so as to formulate a semantic network. SimRank is a versatile similarity measure that evaluates the similarity of objects based on their relationships with other objects in a structural context using graphical model it is formulated as Eq. (1)

$$s(a, b) = \frac{c}{|I(a)||I(b)|} \sum_{j \in I(a)} \sum_{i \in I(b)} s(i, j) \tag{1}$$

$I(a), I(b)$ denotes the in-neighbor of nodes A and B respectively. If $A = B$ then $s(A, B)$ $= 1$. The formalized semantic network as well as the informative term obtained through enrichment of query words are subjected to computation of semantic similarity using the Pianka, Lance and William index. The niche overlap index of Pianka is averaged across each unique species pair x and y. The index is symmetric, with a normalization term for the overlap between species x and y in the denominator as described in Eq. (2).

$$\frac{\sum_i x_i y_i}{\sqrt{(\sum_i x_i^2)(\sum_i y_i^2)}} \tag{2}$$

whereas the Lance and William index is calculated as in Eq. (3) for species x and y.

$$1 - \frac{1}{n} \sum_i \frac{|x_i - y_i|}{(x_i + y_i)} \tag{3}$$

Both are set to step deviation of 0.35, a stringent step division is not used as more number of entities are to be harnessed as this is not the final step. The model is then subjected to the harmony search as two different objective function are used with the same deviants and also because of the harmony search optimization algorithm, Piyanka and Lance & William index is not set to very stringent step deviation. The Harmony Search Algorithm is a metaheuristic optimization algorithm that was inspired by the musical improvisation process. It employs a population-based approach to find optimal solutions by simulating the process of creating harmony, integrating exploration and exploitation. It is used so the instance solutions set is transformed into a final optimal solution set of entities and informative terms which are more relevant. The data set is classified by randomly extracting the entities from the semantic network and features dataset is classified using the XGBoost classifier. XG boost is not a deep learning classifier but is a strong machine learning classifier and uses gradient boosting to construct a robust predictive model. It combines multiple poor learners, typically decision trees, to rectify past errors and achieve high precision. It also effectively manages structured data and prevents overfitting via regularization techniques and gradient-based optimization because the features are to be converged and not to be deviated. The classified instances are subject to semantic similarity computation with that of the optimal set of informative terms using the Adaptive PMI measure and Lance and William index. Pointwise mutual information (PMI) quantifies the chance of two words occurring by assuming that single word frequency causes it. The algorithm calculates the log of the likelihood of co-occurrence of the words (a and b) divided by the product of individual probabilities. The formula can be depicted as Eq. (4).

$$PMI(a, b) = \log(\frac{p(a, b)}{p(a)p(b)}) \tag{4}$$

Normalisation of PMI can be done by dividing PMI by either $- p(a)$, $- p(b)$ or $- p(a, b)$. Adaptive PMI is formulated as Eq. (5).

$$APMI(a, b) = \frac{PMI(a, b)}{p(a)p(b)} + y \tag{5}$$

The Adaptive Pointwise Mutual Information measure computes semantic similarity by improving the PMI measure. The adaptive coefficient y makes the APMI better than the Normalized PMI approach it also boosts semantic heterogeneity computation confidence and word relevance. APMI measure is said to a step deviation of 0.65, whereas the Lance and William measure is said to a threshold of 0.35. Reason for relaxing the threshold is because already the optimal set of terms has been calculated. Finally, all the entities are arranged in the order of increasing semantic similarity. Open link is created so it generates a facets map. As this said map is yielded all the images relevant to this facet are also arranged in the same increasing order of similarity and is used for recommending. The process is continued until the user is satisfied.

4 Results and Evaluation

Experimentations are conducted on 4 distinct independent data sets for web image annotation. Vegetable Image Dataset(https://www.kaggle.com/datasets/misrakahmed/vegetable-image-dataset), House Rooms & Streets Image Dataset(https://www.kaggle.com/datasets/mikhailma/house-rooms-streets-image-dataset), Landmark Image Dataset(https://ig.shaip.com/offerings/computer-vision-data-catalog), Vine Trunk Image/Annotation Dataset (http://doi.org/10.5281/zenodo.5362354). These data set a strategically integrated by annotating the categories using customized annotator. And also by feeding the categorical annotations from the image labels itself. Records are reading based on the annotation similarity the overall dataset is synthesized. Also using a customized crawler several images which are labelled are loaded on the current functioning web 3.0 in order to populate the image space as well. So a image data set comprising of a rich image space and rich term space is obtained on which experimentations are done. Potential performance metrics for the subsequent MLSM model include Precision recall accuracy f-measure percentages and false Discovery rate. The proposed MLSM model offers the highest Precision 94.09%, highest recall 96.91%, and highest accuracy 95.5% and f measure of 95.47% accompanied by lowest FDR of 0.06%. In order to evaluate the efficacy of the proposed model, it is compared to baseline models IIT, MIT, and PSIR. IIT and MIT have developed a tag recommendation model for images, but these categories are ultimately used to recommend images, making tag recommendation an intermediate step in image recommendation. The proposed MLSM model has the same number of queries as the baseline models and is executed in the same environment. PSIR model has yielded Precision of 87.32%, recall of 90.17%, accuracy of 88.74% and F-measure of 88.72% also IIT model produced Precision of 89.45%, recall of 91.47%, accuracy of 90.46% and F-measure of 90.44% and the MIT model had Precision of 90.11%, recall of 92.66%, accuracy of 91.38%, and F-measure of 91.37%. As a knowledge-centered model, the proposed model produces the highest Precision recall accuracy and f measure. In which the user's query is processed using topic modelling in the form of LDA,

WIKI data NELL, and YAGO knowledge stores, which are further enhanced with the bag of world model from an external corpus to produce informative terms. The dataset is also classified using machine learning and deep learning models, such as XG-boost and Bi-LSTM, with features from the constructed semantic network. In addition to the use of strong similarity measures, the model outperforms all other baselines. The reason why PSIR cannot perform better is because it is an image recommendation model that incorporates personalization and social relevance. A bipartite graph is constructed based on the correlation between users. However, the categories on the social website are extremely limited and do not provide a robust ecosystem of supplementary knowledge. In the model, knowledge growth is absent and limited, so the model must be improvised by incorporating a strong relevance computation mechanism and increasing the quantity of auxiliary knowledge. IIT model is a tag recommendation model, and images are returned based on the tag in the proposed Framework for comparing results. Despite using the image context, the model does not perform better because the tags are retrieved from Flickr, the social media platform. It has completely insufficient identifiers. Visual similarity and text statistics are employed, both of which are robust measures, but the model's efficacy could be improved because of the scarcity of identifiers. In addition, the MIT model does not perform as anticipated despite utilizing map-based tag recommendation and maximal a posteriori estimation. Utilizing a visual folksonomy that serves as metadata for informal social classification. However, classifying and managing metadata and making it permeable to the model are insufficient and lack diversity. In addition, a limited relevance computation measure prevents the model from performing better (Table 1).

Table 1. Comparison of Performance of the proposed MLSM with other approaches

Model	Average Precision %	Average Recall %	Average Accuracy %	Average F-Measure %	FDR
PSIR [7]	87.32	90.17	88.74	88.72	0.13
IIT [8]	89.45	91.47	90.46	90.44	0.11
MIT [9]	90.11	92.66	91.38	91.37	0.10
Proposed MLSM	94.09	96.91	95.5	95.47	0.06

From Fig. 3, recall percentage versus recommendation is depicted. MLSM occupies the highest in the hierarchy followed by MIT followed by IIT and last is the PSIR. Highest performance of MLSM can be contributed to strong relevance computation measures and a dense auxiliary knowledge.

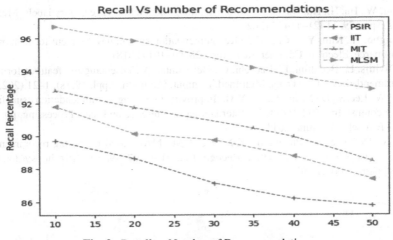

Fig. 3. Recall vs Number of Recommendations

5 Conclusion

Proposed MLSM model is a knowledge driven framework where incremental aggregation of knowledge is used to enhance the information content of both the query and data set driven terms is achieved. Metadata generation increases the density of data quite exponentially and bi-LSTM model is used to classify the metadata. Data set is classified using the XG Boost classifier, and strong semantic similarity measures such as Priyanka index, Lance and Williams index, SimRank, APMI are used in various facets and stages in the network. Knowledge from standard repositories helps in enrichment of the query terms. LDA is also used to contextualize the data set through the names and categories extracted. Facet map are finally generated to rank and recommend images. Model produces best overall accuracy of 95.5% and f-measure of 95.47% land lowest FDR of 0.06. We could enhance the proposed framework by reducing the learning load into the model and increasing the strength of the semantic related-ness measures. Learning load can also be reduced by using strategies such as gamification or nature inspired techniques in order to replace the learning strategies to explainable AI based strategies.

References

1. Shimizu, R., Saito, Y., Matsutani, M., Goto, M.: Fashion intelligence system: an outfit interpretation utilizing images and rich abstract tags. Expert Syst. Appl. **213**, 119167 (2023)
2. Hyvönen, E., Saarela, S., Styrman, A., Viljanen, K.: Ontology-based image retrieval. In: WWW (posters) (2003)
3. Gao, S., Wang, Z., Chia, L.T., Tsang, I.W.H.: Automatic image tagging via category label and web data. In: Proceedings of the 18th ACM International Conference on Multimedia, pp. 1115–1118 (2010)
4. Rawat, Y.S., Kankanhalli, M.S.: ConTagNet: exploiting user context for image tag recommendation. In: Proceedings of the 24th ACM International Conference on Multimedia, pp. 1102–1106 (2016)

5. Zhang, W., Hu, H., Hu, H., Yu, J.: Automatic image annotation via category labels. Multimed. Tools Appl. **79**, 11421–11435 (2020)

6. Hwangbo, H., Kim, Y.S., Cha, K.J.: Recommendation system development for fashion retail e-commerce. Electron. Commer. Res. Appl. **28**, 94–101 (2018)

7. De Divitiis, L., Becattini, F., Baecchi, C., Del Bimbo, A.: Disentangling features for fashion recommendation. ACM Trans. Multimed. Comput. Commun. Appl. **19**(1s), 1–21 (2023)

8. Eom, W., Lee, S., De Neve, W., Ro, Y.M.: Improving image tag recommendation using favorite image context. In: 2011 18th IEEE International Conference on Image Processing, pp. 2445–2448. Brussels, Belgium (2011)

9. Lee, S., De Neve, W., Plataniotis, K.N., Ro, Y.M.: MAP-based image tag recommendation using a visual folksonomy. Pattern Recogn. Let. **31**(9), 976–982 (2010). https://doi.org/10.1016/j.patrec.2009.12.024

Protocol Security in 6th Generation (6G) Networks

Tanya Garg[1]($\boxtimes$)(iD), Gurjinder Kaur[2], and Prashant Singh Rana[1]

[1] Thapar Institute of Engineering and Technology, Patiala, Punjab, India
{tanya.garg,prashant.singh}@thapar.edu
[2] Sant Longowal Institute of Engineering and Technology, Longowal, Punjab, India

Abstract. 6th generation networks are expected to be there most probably in year 2030. It is believed that their existence will enable the technology of Internet of Everything. People are already using the 5G Networks and facing several issues in that due to which the 6G networks are very important to develop. 6G technology ensures to connect every device present in the world with the internet providing Quality of Service. The 6G networks are still in discussion stage and it is identified that there is need of tackling the several issues related to security and implementation to get better and secure services from 6G Network. A lot of research is being carried out to ensure the security and integrity of 6G Technology. Various issues need to be addressed before implementing 6G Networks. The security is the main aspect behind the 6G Networks so security must be ensured at each layer of network model. The protocol security is very crucial issue to be considered. In this article, the security is the major concern that has been discussed in detail. Very few researchers have done the work on 6G protocol security in last three years. So, in this article, the main focus is to explore the security protocols and the issues related with the protocols to provide 6G network security have been elaborated to provide the way to the researchers to explore and resolve the 6G Security protocols. The future issues and challenges that need to be focused have also been discussed in this work.

Keywords: 6G (6th generation) · Internet of Things (IoT) · Security · Protocols · Wireless Communication · 5G (5th generation)

1 Introduction

5th generation (5G) organizations are adequate for the current applications, notwithstanding, it has a few constraints and can't uphold future applications. Hologram exchanges, for instance, might need information rate that is usually measured in terms of TBps, whereas 5G supports a top speed of 10 Gbps. The devices to be connected have to be increase for the communication among them and the host has significantly increased as a result of the IoT ongoing growth. As a result, 5G will probably be unable to provide network availability to many devices and the expected future clients will have to face the issues of congestion

R. Muthalagu et al. (Eds.): CINS 2023, CCIS 1978, pp. 47–62, 2024.
https://doi.org/10.1007/978-3-031-48984-6_5

in network, The networks are upgraded an designed physically in the current scenario that is not capable to scale up and down to work in larger organizations. The future organisations that are expected to use 6G networks on combining with AI (Artificial Intelligence) technologies are expected to overcome these limits [1].

The fundamental guidelines and conventions that control data transfer, communication, and interactions inside the network are these protocols. New and better protocols are being created as 6G networks expand to meet the increasing demands of a hyper-connected world. We will examine some of the major advancements in this field and look into the significance of protocols in 6G in this article.

While the 5G organization has recently begun to carry out industrially, however its effects on the economy, and society are as yet unclear. Number of researchers across the globe have begun investigating 6G correspondence organizations and give their hypotheses and structures. Researchers investigated that what are the new components in 6G and broke down administrations and applications where it can assume a crucial part. In any case, there exists no inside and out study which investigates every 6G idea, administrations, AI coordination, engineering, a scientific categorization for AI strategies in 6G organizations, genuine use case, and future situations. With an expanded correspondence prerequisite of end-clients and network administrators, the QoS, QoE, and to accomplish more dependable wireless correspondence network which is the primary explanation for evolution in the correspondence channel from original (1G) to fifth-age (5G) and is currently exchanging towards the 6th generation (6G) organizations. With the quick implementation of advance technologies, remote information traffic has definitely expanded, and current cell organizations (indeed, even the approaching 5G) can't totally match the rapidly rising specialized necessities. To address the approaching troubles, the sixth era (6G) flexible association is depended upon to extend the high particular standard of new reach and energy-capable transmission strategies [2]. Nowadays, 6G development is one of the most developing field of exploration nowadays. Besides, the 6G will be shown as an unmistakable benefit perspective in various fields. Hence, the visioning of the 6G development means a lot to change the state of the art world. The present moment, various countries are conveying the 5G development going for it.

However, the Internet of Everything's (IoE) absolute requirements cannot be met by 5G or Beyond 5G (B5G) (IoE). As a result, 6G has gained notoriety. The problems with B5G convenient correspondence are now being attended to by experts. It is expected that 6G will have an impact on a few fields that need to be visioned. However, there are various issues with 6G, so in this section, we explore all of these issues and difficulties from every angle. [3]. The organization of this article is as shown in the figure below: Initially, the 6G technology has been introduced, after that the need of security for 6G network has been discussed, Then in Sect. 3, Security issues and challenges have been described in detail and in Sect. 4, the work that has been already carried out by various researchers for the security of 6G technology has been described in detail. Finally, Sect. 5 is focused on the protocols that van be used for 6G networks (Fig. 1).

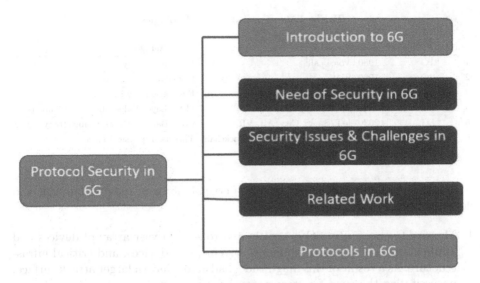

Fig. 1. Article Organization

2 Need of Security in 6G

The difficulties in making a reliable 6G are multidisciplinary spreading over innovation, guideline, techno-financial aspects, governmental issues and morals. A blend of the current guideline, monetary motivations and innovation are keeping up with the current degree of hacking, absence of trust, protection and security on the Internet. In 6G, this won't get the job done, in light of the fact that actual security will increasingly more rely upon data innovation and the organizations we use for correspondence. This is the reason of need for 6G reliable networks. The concept of trust, protection and security are interconnected for these networks when implemented inside organizations [4].

The Sixth Generation Networks (6G) will be utilized for the fabrication of huge world limit for detecting the world, getting it and programming it. As an outcome, notwithstanding loss of data, loss of authority over your gadget or host or deficiency of cash, break of data security which can lead to loss of confidential data and privacy. To say the least, during global struggles unfamiliar cyberwar troops could cause destruction in a country on a level that utilizing customary fighting won't be expected to pressure the casualty to acknowledge the agreements gave by the assailant. Like we are finding in the 5G period, public safety concerns are assuming an expanding part in versatile innovation. This pattern will be much more unmistakable in 6G. To address these worries, 6G organization should uphold security and trust with the end goal that the subsequent degree of data security in 6G and the bundle information networks where 6G gives availability to is fundamentally better compared to in cutting edge networks that are mostly used today. [5]. The various technologies are compared and summarized in the table below (Table 1):

Table 1. Comparison of Technologies

Sr.No.	Technology	Year	Focus	Disadvantages
1	1G	1980	Voice call	Lack of Security
2	2G	1990	SMS	One-way authentication
3	3G	2000	Web Applications	Encryption key issues
4	4G	2010	Mobile Applications	MAC Layer Vulnerabilities & Threats
5	5G	2020	IoT/Smart City/AR, VR	Cost, Battery Drain, Connectivity issues
6	6G	2030	Autonomous Systems, Blockchain	Threats in physical layer

For a number of reasons, security is a crucial factor in the development and deployment of 6G networks.

1. Massive Connectivity: 6G is anticipated to serve a vast array of devices and applications, including autonomous systems, IoT devices, and critical infrastructure. As a result of this magnitude, bad actors have a larger attack surface, necessitating the need for strong security measures.
2. High Data Rates: 6G networks will provide noticeably higher data rates, which may allow for the transmission of more sensitive and priceless data. Because of the speedier data transmission, cybercriminals looking to eavesdrop or interfere with these connections may be attracted.
3. Mission-critical applications, such as remote surgery, autonomous vehicles, and smart city management, are expected to be made possible by 6G. Any security lapses in these applications may result in grave or widespread repercussions.
4. Privacy Issues: As data gathering and sharing increase, privacy issues become more and more important. Huge volumes of sensitive and private data will be handled by 6G networks, making user privacy protection essential.
5. Threats from Quantum Computing: As quantum computing develops, it poses a serious risk to traditional encryption algorithms. To protect against future assaults, 6G networks will need to implement encryption techniques that are resistant to quantum computing.
6. Network Slicing: A key component of 6G, network slicing enables the construction of virtual networks with unique properties. Strong security measures must be in place on each slice to guard against unauthorized access or intervention.
7. Edge Computing: 6G will mainly rely on edge computing to lower latency and increase efficiency. This distributed design needs secure edge devices and communication since it expands the possible attack surface.
8. AI and machine learning: If 6G systems and protocols are not properly secured, they may be exploited. Network operations could be affected by adversarial assaults on AI models or decision-making techniques.
9. Supply Chain Security: There are potential for hardware and software manipulation due to the worldwide nature of technology supply chains. The integrity of parts and software must be guaranteed for 6G security.

3   Security Issues and Challenges in 6G

The 6G network will be an integration of advanced technologies like the Artificial Intelligence, the Internet of Things, the blockchain and data mining techniques along with edge computing to optimize the network performance. The aim is to provide full space coverage as compared to 5G i.e. undersea as well as ground level network coverage. With these improvements it is also important to reduce security and privacy issues at different levels of implementations (Fig. 2).

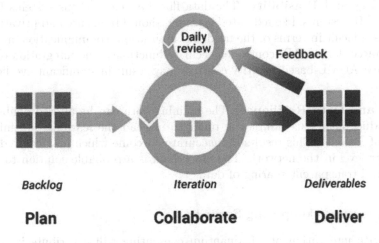

Fig. 2. Security in 6G

With the advancements in the network architecture the security flaws have also become more prominent. Also, data security is a crucial issue to be resolved in the 6G edge networks. There are few security issues and challenges in 6G like how can the integrity and confidentiality can be maintained without affecting user experience or detecting threats in a heterogeneous dense network environment or the level of security provided by extremely novel technologies like artificial intelligence and quantum computing. Some of the 6G security related aspects have been describe here in this section.

3.1   Artificial Intelligence Security

This advance nature in networks can only be achieved by machine learning techniques and algorithms [1]. Other key features of 6G networks include assured Quality of Service of IoT devices, absolute connectivity and processing huge volumes of data. All of these features can be accomplished in a 6G network with the help of Artificial Intelligence [1] AI has the ability to learn, analyze, classify and recognize distinct data patterns and provide complex decision making abilities. Therefore, the application and integration of AI/ML in 6G networks provide better performance. The other aspect to the integration of AI/ML with

6G networks is providing a secure environment with robust performance. Few security issues in this domain are described as follows.

Trustworthiness. In a 6G network major network functions like security are controlled by the AI component. At the same time the trustworthiness of AI in terms of security is still questionable. Hence, tools like integrity checks, formal verification techniques and trusted computing enablers are vital and necessary.

Scalability and Feasibility. The data flow between various systems for distributed ML systems, like federated learning, should be secure and private. Scalability is difficult in terms of the needed computation, communication, and storage resources for AI/ML controlled security functions. The integration of these flows with AI/ML-based security controls may result in significant overhead.

Model and Data Resilience. The training data can be altered deliberately to introduce malicious samples in order to mislead the learning and inference phases of a model. This results in inaccurate outcome which may cause damage at higher level in the network. The blockchain is a probable solution to ensure secure and transparent sharing of data.

3.2 Quantum Computing Security

Along with integration of AI, quantum computing will contribute in efficient implementation of 6G networks. Quantum computing is an advance technology that uses the quantum nature of information. This will introduce novel ways of cryptography in communication channels which will provide more randomness and security. The recent research and advancements in quantum computing aided communication channels indicate that the present day channels can be replaced with noiseless communication channels. As 6G networks will provide full space communication, applications of quantum computing will assist in long distance communications. Therefore, quantum computing will contribute in detection, reduction and prevention of any security related vulnerabilities [6].

Currently there are promising researches regarding combining machine learning based security along with quantum encryption techniques for more reliable communication links in 6G networks. An important feature of quantum communication is that if there is any malicious activity like eavesdropping or replication of data in the communication link this will affect the quantum state. The receiver will notice this change and know immediately of some malicious activity. Therefore, it is easy to know about any interference in a quantum communication channel. There are advancements in the arena of developing quantum cryptography but certain aspects such as fiber attenuation and operation errors need to be considered for long distance communication. A number of techniques like quantum dense coding, quantum key distribution (QKD), quantum secret sharing, etc. are required to be orderly and accurately incorporated in quantum communication channel to make it fully secure [3] Another major concern is redesigning some of the important protocols related to cryptography algorithms for an era of quantum cryptography algorithms. There might be concerns regarding the complexity of the new cryptography algorithms and the size of the keys for low-power and low-cost devices. Hence, there might be some unexplored and unforeseen challenges in securing 6G networks using quantum cryptography algorithms (Fig. 3).

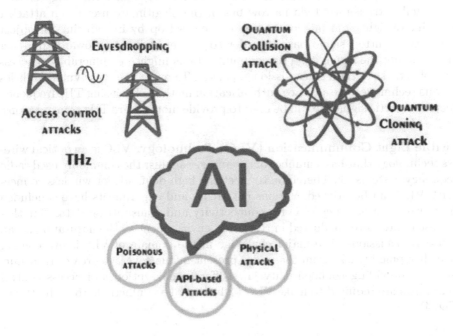

Fig. 3. Quantum Computing security

3.3 Physical Layer Security

There are basically three layers in the network: Physical Layer, Network Layer and Application Layer. The Physical layer is considered to be the first line of defense that performs the task of providing security to the nodes that are very complex. It can provide an additional layer of security by using techniques like signal processing, artificial noise injection, key generations, etc. Technologies like Terahertz (THz), Visible Light communication and Molecular Communication is discussed further.

TeraHertz (THz) Technology. TeraHertz technology or THz is a crucial contributor to 6G technology. At present the Radio Frequency band is almost fully occupied and the mm-wave band (currently being used for 5G technologies) will not suffice really high transmission rates in 6G technology. Tetra hertz communication utilizes 0.1–10 THz band. These frequencies make use of both electromagnetic waves and light waves and provide a secure communication channel [3]. Also, they use narrow beam and short pulse duration which reduces chances of eavesdropping to a minimal hence, increasing security. For instance, when a signal is transmitted for an authorized receiver it is difficult for an unauthorized user to be on the same signal path. At the same time THz technology also has few security and privacy issues like there are few researches that suggest that if a signal is transmitted via narrow beam, the illegitimate user or an attacker is in line-of-sight transmission he/she can eavesdrop by introducing an object in the transmitted signal path in order to scatter radiations towards him/her and intercept the signal [6]. THz communications might be vulnerable to access control attacks or data transmission exposure. To overcome these vulnerabilities various techniques like advance authentication methods like using THz frequency electromagnetic signature can be used to provide more secure THz transmissions.

Visible Light Communication (VLC) Technology. VLC is an optical wireless technology that has a number of advantages against the commonly used radio frequency systems [6]. Therefore, to meet the high demands for wireless connectivity VLC can be utilized. Various researches and experiments have concluded the advantages of using VLC for connectivity and transmitting data. But theses researches were conducted in restricted environment as the natural light can affect transmissions. The major challenge in the domain of VLS is overcoming eavesdropping attacks from the nodes present in the coverage area of transmitters and providing confidentiality [4]. Different encryption and access control techniques are required to make VLC technology less vulnerable to such attacks (Fig. 4).

Molecular Communication. Molecular communications is inspired by the nano scalar entities where bio-nanomachines communicate among each other using molecules or biochemical signals in aqueous environments. It is a result of advancements in nanotechnology and bioengineering combined with synthetic biology. The major advantage of molecular communication is that the devices involved are of micro and nano scale and consume less energy as compared to other devices. Hence, it is an emerging technology for 6G communications.

Due to its novel and sophisticated nature molecular communication has attracted researchers from multiple domains and conducted various experiments. These experiments have highlighted the major security and privacy issues related to using molecular communication in transmitting information. The issues primarily revolve around authentication and encryption while communication between sender and receiver. This technology is also prone to flooding attacks, jamming or de-synchronization at different levels of communication.[14] At the same time researches are being conducted to provide more security and reliability in molecular communication. Researchers have suggested a unique coding scheme to increase security or a different key distribution proposal for

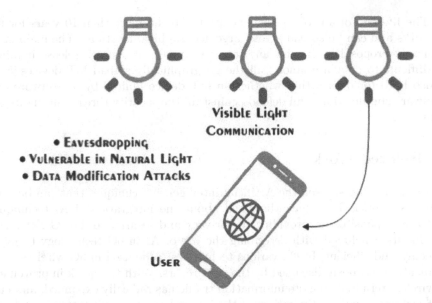

• Eavesdropping
• Vulnerable in Natural Light
• Data Modification Attacks

Visible Light
Communication

User

Fig. 4. Visible Light Communication Technology

molecular communications. The main aim of any such study is to provide more techniques to secure molecular communication by developing advance techniques and make molecular communication more reliable for 6G networks.

3.4 IoT Device Security

IoT devices will be an integral part of 6G network where there will be more than 125 billion devices by 2030 [5]. IoT devices introduce a number of challenges in providing information security and confidentiality. For example, just like the 2016 Mirai attack an aggregate of small IoT devices in a huge network can cause severe damage. Another challenge posed by IoT devices are that for an advanced network just like 6G low-end devices will not be suitable [5]. Also, more reliable devices that are in accordance to advance security techniques, high computing power and energy constraints need to be considered and used.

Presently, there are a substantial security issues and challenges at IoT device level. Primarily, IoT devices need to be more secure and proficient individually in terms of hardware and software. New techniques and algorithms for IoT devices need to be intended to provide lightweight cryptography in post quantum world. There are various attacks in mobile networks that have not been explored yet. In order to avoid any kind of damage from such attacks techniques to prevent such attacks need to be devised and considered. Also, it is important to introduce decentralized anomaly detection in networks as large as 6G. Hence, it is important to make the IoT devices secure and efficient but by using techniques that contribute to faster yet reliable 6G network.

The lifespan of an IoT device is expected to be more than 10 years for 6G networks and can be spread across large geographical locations. The main issue with this proposal is that for any security update or a patch release it might be difficult to circulate among all the geographically spread IoT devices.[5] It is also hard to confirm that whether an IoT device limited by its software and hardware specifications will defend against all the security threats during its life span

4   Related Work

Sheth et al., [1] discussed the Artificial intelligence techniques that can be used in 6G Networks. They have discussed about the integration of AI techniques that are responsible for providing the privacy and security to the 6G Networks. and finally concluded with describing the use of AI in 6G technology to make it secure and efficient. In [2], various techniques to be used in 6G wireless communication has been discussed by the researchers. With the quick improvement of wireless terminals remote information traffic has radically expanded, and current cell organizations (indeed, even the impending 5G) can't totally match the rapidly rising specialized prerequisites. The 6th generation (6G) versatile organisation is entrusted with the task of projecting the high specialised norm of new range and energy-productive transmission procedures to deal with the impending challenges. They outlined the most recent research on the exciting 6G development strategies and sketched the probable prerequisites in this article, which has recently attracted a lot of attention. With regard to the 6G Network, numerous issues have been thoroughly discussed, including design issues, security and privacy-related issues.

Chorti et al., specifically focused on physical layer security in 6G networks. They introduced the context aware approach for physical layer security in 6G networks. They described that the actual layer security arrangements arise as serious possibility for low intricacy, low-delay furthermore, low-impression, versatile, adaptable and setting mindful security plans, utilizing the actual layer of the interchanges in truly cross-layer conventions. Shahraki et al., [7] have done the survey of 6G Networks, its applications, utilizing technologies and future challenges that it needs to tackle. More focus was on Cellular IoT. In [8] the researchers focused on architectural changes that need to be done in 5G architectural design for designing the architecture of 6G networks. They examined the related potential advances that are useful in framing, reasonable and socially consistent organizations, enveloping terahertz and apparent light correspondence, new correspondence worldview and block chain. They emphasized on green 6G networks. Soderi, S.(2020), [9] proposed a visible light technology to provide security in 6G networks particularly for physical layer. They discussed about working on the privacy of the upcoming age of remote interchanges by utilizing the watermark-based visually impaired actual layer security in Visible Light Communications (VLCs). Since the development of remote applications and administration, the interest for a safe and quick information move association requires new innovation arrangements fit to guarantee the best counter-measure against security attacks. VLC is perhaps the most encouraging new remote correspondence innovation, because of the chance of utilizing natural fake lights as information move divert in free-space. Then again, VLCs are even innately defenseless to listening in assaults. This work proposed an imaginative plan where red, green, blue (RGB), light-discharging diodes (LEDs) and three shading tuned photograph diodes (PDs) are utilized to get a VLC by utilizing a sticking recipient related to the spread watermarking procedure, this is the main work that deals with physical layer security using VLC by utilizing RGB LEDs.

Long et al., in [10] has done the survey on 5G and 6G network security issues and challenges. In this article, innovation of programming definition organiza-tions (SDN) of 5G and 6G, including framework design, asset the executives, portability the executives, impedance the board, difficulties, and open issues have been reviewed [11]. As a matter of first importance, the framework models of 5G furthermore, 6G portable organizations are presented dependent on SDN advancements. Then, at that point commonplace SDN-5G/6G application situa-tions and central questions are examined. We likewise center around portability the executives approaches in versatile organizations. Also, three sorts of versa-tility the board system in programming characterized 5G/6G are depicted. In [12], 6G remote organizations developed 5G by further expanding dependability, accelerating the organizations and expanding the available transfer speed. These developmental upgrades, along with various progressive enhancements, for exam-ple, high accuracy 3D confinement, super high dependability and outrageous versatility, present another age of 6G-local applications. Such application can be founded on, for instance, circulated, pervasive Artificial Intelligence (AI) and super solid, low-dormancy Internet of Things (IoT). Along with the upgraded availability and novel applications, protection and security of the organizations and the applications should be guaranteed. Conveyed record advancements, for example, blockchain give one answer for application security and protection, however presented their own arrangement of safety and security hazards. In this work, chances and difficulties identified with blockchain utilization in 6G, and map out potential bearings for surpassing the difficulties have been examined.

Kantola, R. [13] has analysed various available frameworks to provide trusted network using 6G technology and finally decided that various threats and attacks needs to be handled to provide trusted framework for 6G networks. You et al., in [14], the drawbacks of 5G technology has been discussed in detail and the way through which 6G technology can resolve all the flaws of 5G have been described. The four Paradigms of 6G network have been explained. The 6G net-works are expected to have global coverage for every type of vehicles like UAV (Unmanned Aerial Vehicle) and terrestrial satellites. Second, all spectra will be completely investigated to additional expansion information rates and associa-tion thickness, including the sub-6 GHz, millimeter wave (mmWave), terahertz (THz), also, optical recurrence groups. Third, confronting the huge datasets cre-ated by the utilization of very heterogeneous networks, various correspondence situations, huge quantities of radio wires, wide data transfer capacities, and new assistance necessities, 6G organizations will empower another scope of brilliant applications with the guide of Artificial intelligence and huge information advances. Fourth, network security should be reinforced when creating 6G networks. This article gives a thorough overview of ongoing advances and future patterns in these four perspectives. Obviously, 6G with extra specialized prerequisites past those of 5G will empower quicker and further interchanges to the degree that the limit among physical and digital universes vanishes.

In [6] Wang et al., has explored the concerned issue of security & privacy for 6G Networks. The reason behind the development of 6G Technology is the drawback of 5G networks. Initially they discussed the existing technologies and their positive/Negative impacts. After that, they discussed the facts that were identified by the existing technologies about the 6G networks. In [15]Kantola, R. (2019) has described the need of providing the trust by the 6G networks for the devices connected. The paper centers around standards and just concerned to some check all together not to mess the conversation with specialized detail. Gui et al., in [16] has depicted the five major services to be provided by the 6G networks considering eight key performance indicators (KPI). The architecture for 6G network has also been proposed and it was analyzed using four different application areas. The various challenges along with their possible solutions have also been stated in this article.Future open areas have also been inspired in the area of 6G. Richard et al., in [17] discussed the need of 6G networks. In this paper, a complete review on 6G-empowered IoT has been presented. The drivers and prerequisites by summing up the arising IoT-empowered applications and the relating prerequisites, alongside the impediments of 5G has ben discussed. Second, dreams of 6G are given as far as center specialized prerequisites, use cases, and patterns. Third, another organization design gave by 6G to empower gigantic IoT is presented, i.e., space-air-ground/underwater/ocean networks improved by edge figuring. Fourth, some advancement advances, for example, AI and blockchain, in 6G have been presented, where the inspirations, applications, and open issues of these advancements for huge IoT are summed up. The utilization instance of completely self-governing driving has been introduced to show 6G support for IoT.

Alsharif et al., in [18] has discussed the vision, innovation and challenges for 6G networks. The key elements of this investigation, which started with a dream, indicated that future 6G would be encouraged in the following measurements: energy effectiveness, insight, phantom productivity, security, mystery, and protection; reasonableness; and customization. After that, we discussed the few potential issues with 6G innovation and the anticipated solutions to support future 6G. At long last, this work closed with worldwide examination exercises that expect to make a dream for future 6G. Lu Y., in [19] has done the survey for security issues of 6G networks and concluded how these issues can be resolved. To this end, the security difficulties of 6G organizations have been examined top to bottom, the essential standards and key advances for managing 6G are clarified according to the point of view of security, and the difficulties that 6G security might look in what's to come are examined. Akhtar et al., in [20] discussed about the requirements and need to shift to 6G networks. Moreover, they discussed about the future research challenges in the field of 6G Networks. They described the five key elements of 6G networks that are as shown in the figure below (Fig. 5):

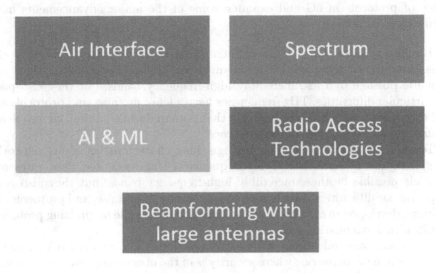

Fig. 5. Components of 6G

5   Protocols in 6G

The protocols should be designed specifically for 6G networks. The protocols must be capable to support 3D and should be able to propagate in different application areas. The protocols for 6G must support heterogeneous devices having different mobility structures. The 6G protocols must be capable to capture the information and easily adaptable to different environments [21]. The protocols to be designed for 6G Networks must support the technologies that are not supported by 5G networks and they must support cloud applications to manage huge amount of data as suggested by European Researchers [22]. In 6G, there can be four layers in protocol stack that are: Physical Layer, Radio link control (RLC), Packet Data Convergence Protocol (PDCP), Radio Resource Control (RRC), and Non Access Stratum (NAS). The protocols to be used by 6G networks can be TCP (Transmission Control Protocol), Internetworking Protocol (IP), PDCP (Packet Data Convergence protocol), SCTP (Stream Control Transmission Protocol), Service Data Adaptation Protocol (SDAP) and User Datagram Protocol. Other protocols can also be used depending upon the scenario.

Protocols will act as the structural foundation of the connected world in the 6G future, where data flows easily, gadgets converse shrewdly, and security is of the utmost importance. For 6G networks to reach their full potential, these protocols must be developed and standardized since they will dictate how objects, software, and services interact inside this cutting-edge wireless ecosystem. The development of 6G protocols by scientists and engineers paves the way for a future that is more connected, secure, and technologically advanced. The importance of protocols in 6G and examine some of the major advancements have been discussed below:

1. Terahertz Bands: One of the distinguishing characteristics of 6G networks is their use of terahertz (THz) frequencies. Unprecedented data rates are made possible by these incredibly high-frequency bands, but they also pose particular difficulties. THz frequencies have a lot of promise, and protocols are being developed to effectively utilize that potential while minimizing problems like signal attenuation and interference.
2. Terahertz Bands: One of the distinguishing characteristics of 6G networks is their use of terahertz (THz) frequencies. Unprecedented data rates are made possible by these incredibly high-frequency bands, but they also pose particular difficulties. THz frequencies have a lot of promise, and protocols are being developed to effectively utilize that potential while minimizing problems like signal attenuation and interference.
3. Quantum communication protocols are positioned to play a key role in securing data in 6G networks, where security is of the utmost importance. To counteract the risks posed by quantum computing, quantum-resistant encryption techniques are being developed, preserving the privacy and integrity of data.
4. AI-Driven Protocols: A key element of 6G networks is artificial intelligence (AI) and machine learning (ML). Protocols that leverage AI can optimize network resource allocation, predict network congestion, and enhance quality of service. These smart protocols are expected to make networks more adaptive and efficient.
5. Holographic Beam forming: To achieve the massive connectivity promised by 6G, advanced beam forming techniques are essential. Holographic beam forming protocols enable precise and dynamic control of multiple beams, enhancing network coverage and capacity.
6. Tactile Internet Protocols: Beyond traditional data transmission, 6G aims to support tactile internet applications that require ultra-low latency. Protocols designed for tactile internet will enable real-time interactions for applications like remote surgery and haptic feedback systems.
7. Green Protocols: Sustainability is a growing concern, and 6G networks aim to be more energy-efficient. Green protocols are being developed to optimize power consumption, reduce carbon footprints, and support Eco-friendly network operations.
8. Security-First Protocols: With the proliferation of cyber threats, security is a top priority in 6G. Protocols are being designed with a security-first approach, integrating features like end-to-end encryption, authentication, and anomaly detection to protect data and network integrity.

6   Conclusion

In conclusion, the transition from 5G to 6G networks marks a major advancement in wireless communication. While 5G paved the way for improved connection and new opportunities, it also revealed several shortcomings that need to be fixed. These problems could be solved and a wave of revolutionary improvements could be brought about by 6G networks, which are now in the developmental stages. As we anticipate the 6G era, we may picture a world where connectivity is commonplace, the Internet of Things thrives, and real-time experiences are accepted as the standard. To enable the safe and secure implementation of 6G networks, security issues must be thoroughly evaluated and addressed as with any new technology. The main motivation behind creation of 6G networks have been discussed in this article, as well as its critical importance to our way of life. Furthermore, it has brought to light how crucial it is to analyze and address the security risks posed by this developing technology.

Researchers, technologists, and policymakers will need to collaborate in the upcoming years to fully utilize 6G while preventing any security flaws. By working together, we can lay the groundwork for a time when 6G networks will fundamentally alter how we connect, communicate, and interact with the outside world.

References

1. Sheth, K., Patel, K., Shah, H., Tanwar, S., Gupta, R., Kumar, N.: A taxonomy of AI techniques for 6G communication networks. Comput. Commun. **161**, 279–303 (2020)
2. Yang, P., Xiao, Y., Xiao, M., Li, S.: 6G wireless communications: vision and potential techniques. IEEE Netw. **33**(4), 70–75 (2019)
3. Nayak, S., Patgiri, R.: 6G communication: envisioning the key issues and challenges. arXiv preprint arXiv:2004.04024 (2020)
4. Ylianttila, M., et al.: 6G white paper: research challenges for trust, security and privacy. arXiv preprint arXiv:2004.11665 (2020)
5. Chorti, A., et al.:. Context-aware security for 6g wireless the role of physical layer security. arXiv preprint arXiv:2101.01536 (2021)
6. Wang, M., Zhu, T., Zhang, T., Zhang, J., Shui, Yu., Zhou, W.: Security and privacy in 6G networks: new areas and new challenges. Digit. Commun. Netw. **6**(3), 281–291 (2020)
7. Shahraki, A., et al.: A comprehensive survey on 6G networks: applications, core services, enabling technologies, and future challenges. arXiv preprint arXiv:2101.12475 (2021)
8. Tongyi Huang, W., Yang, J.W., Ma, J., Zhang, X., Zhang, D.: A survey on green 6G network: architecture and technologies. IEEE Access **7**, 175758–175768 (2019)
9. Soderi, S.: Enhancing security in 6g visible light communications. In: 2020 2nd 6G wireless summit (6G SUMMIT), pp. 1–5. IEEE (2020)
10. Long, Q., Chen, Y., Zhang, H., Lei, X.: Software defined 5G and 6G networks: a survey. Mob. Netw. Appl. 1–21 (2019)
11. Mucchi, L., et al.: Physical-layer security in 6G networks. IEEE Open J. Commun. Soc. **2**, 1901–1914 (2021)

12. Nguyen, T., Tran, N., Loven, L., Partala, J., Kechadi, M.T., Pirttikangas, S.: Privacy-aware blockchain innovation for 6G: challenges and opportunities. In: 2020 2nd 6G Wireless Summit (6G SUMMIT), pp. 1–5. IEEE (2020)
13. Kantola, R.: Trust networking for beyond 5G and 6G. In: 2020 2nd 6G Wireless Summit (6G SUMMIT), pp. 1–6. IEEE (2020)
14. You, X., et al.: Towards 6G wireless communication networks: vision, enabling technologies, and new paradigm shifts. Sci. Chin Inf. Sci. 64(1), 1–74 (2021)
15. Kantola, R.: 6G network needs to support embedded trust. In: Proceedings of the 14th International Conference on Availability, Reliability and Security, pp. 1–5 (2019)
16. Gui, G., Liu, M., Tang, F., Kato, N., Adachi, F.: 6G: opening new horizons for integration of comfort, security, and intelligence. IEEE Wirel. Commun. 27(5), 126–132 (2020)
17. Guo, F., Yu, F.R., Zhang, H., Li, X., Ji, H., Leung, V.C.: Enabling massive IoT toward 6G: a comprehensive survey. IEEE Internet Things J. 8, 11891–11915 (2021)
18. Alsharif, M.H., Kelechi, A.H., Albreem, M.A., Chaudhry, S.A., Zia, M.S., Kim, S.: Sixth generation (6G) wireless networks: vision, research activities, challenges and potential solutions. Symmetry 12(4), 676 (2020)
19. Yang, L.: Security in 6G: the prospects and the relevant technologies. J. Ind. Integr. Manage. 5(03), 271–289 (2020)
20. Akhtar, M.W., Hassan, S.A., Ghaffar, R., Jung, H., Garg, S., Hossain, M.S.: The shift to 6G communications: vision and requirements. Hum. Centric Comput. Inf. Sci. 10(1), 1–27 (2020)
21. Saad, W., Bennis, M., Chen, M.: A vision of 6G wireless systems: applications, trends, technologies and open research problems. IEEE Netw. 34(3), 134–142 (2019)
22. Bernardos, C.J., Uusitalo, M.A.: European vision for the 6G network ecosystem (2021)

Efficient Method for Video Sentiment Analysis

Shailaja Uke[1] and Nilesh Uke[2]([envelope])

[1] Vishwakarma Institute of Technology, Pune, India
[2] Trinity Academy of Engineering, Pune, India
nilesh.uke@gmail.com

Abstract. Nowadays, the acquisition of deep knowledge is carried out in many fields like tracking objects from image/video, measuring position, acquiring textual and image content, visual value detection, and recognizing hand gestures. Various deep learning models are available based on classification and regression like Convolutional Neural Networks (CNN), decision trees, linear regression which comes under supervised and Self Organizing Map (SOM), Boltzmann Machines, and Autoencoders which comes under unsupervised models based on clustering. For image classification, one of the well-performing model is CNN. It provides good performance when compared to other deep learning models. In this paper, we propose an efficient method that will analyze sentiment from video. This is carried out by using Convolutional Neural Networks (CNN).

The findings are also compared with various well-known deep learning approaches, and the results obtained by the proposed method are higher when compared to existing ones. The proposed methodology gives an accuracy level of 92%. The proposed methodology can be used in various applications such as audio/video data sentiment analysis, monitoring social media, and customer feedback analysis, in the education field to draw student's opinions.

Keywords: Convolutional neural network · Deep learning · Video Sentiment Analysis · Machine learning · Emotion Recognition · Image Classification · Neural Networks · Visual Analytics

1 Introduction

Deep Learning (DL) is a subset of machine learning, which facilitates a computer to work on data the same way the human brain does. An in-depth learning model is introduced to further differentiate information on a structure such as how one can make decisions. In-depth learning employs a horizontal framework of numerous algorithms to accomplish this. In deep learning, we consider neural networks that point to an image based on its characteristics. This is achieved in order to create a complete output model that is able to solve the difficulties faced by traditional methods. The incorporated model extractor ought to be able to master accurately extracting the distinguishing characteristics from the image training data.

Sentiment analysis, commonly referred to as opinion mining, is a branch of natural language processing (NLP) that seeks to recognise and extract subjective information from text data, such as people's attitudes, feelings, and views. Sentiment analysis of videos is becoming more and more important as consumers' emotional responses to various types of video material become more and more well-known. Due to the complicated structure of video data, which includes various modalities, such as visual, audio, and textual information, video sentiment analysis is a difficult process. As well as changing over time, a video's sentiment can also alter depending on the context and the kind of content being delivered. Classifying and categorising things into groups and categories is a systematic process called dividing. By using data-driven computer training, image separation was developed to close the gap between computer vision and human perception. Frames taken from videos can be classified as images by assigning them to a particular category depending on their content [1]. In research, a number of solutions have been put up to deal with these problems, including rule-based approaches, machine learning methods, and deep learning models. Due to their capacity to extract spatial and temporal characteristics from video frames, Convolutional Neural Networks (CNNs) have distinguished themselves as one of the most effective methods for sentiment analysis of videos..

In this research, we suggest an innovative approach for CNN-based video sentiment analysis. The suggested technique collects spatial and temporal characteristics from video frames and recognises sentiment trends over time. A significant video dataset including a variety of emotions and attitudes was used to train the CNN-based algorithm. We assess the effectiveness of our suggested technique using a number of criteria, such as accuracy, precision, recall, and F1-score, and we contrast it with contemporary methodologies. The objective of Video Sentiment Analysis is the automatic al-location and recognition of frames taken from videos to thematic classes of emotion. Unsupervised classification and supervised classification are both different forms of classification. Two parts make up the sentiment analysis process: first, the system must be trained, and subsequently it must be tested using CNN.

2   Related Study

Sentiment analysis is a topic that has received a lot of attention from researchers in the fields of deep learning, recognition of patterns, and interaction between humans and computers. Removing the image's components allows for image separation [2]. The medium-sided learning approaches often concentrate on the coding or merging process, but in this case they emphasise how the procedure used to create the image has a big impact on how elements are removed from the image. Using a sequential picture detection method, the image content analysis was completed successfully for the feature's release. Each image in this case is separated into a number of semantic components, including images of the structure and texture. Using various feature extraction techniques, the structure and texture of a semantic image can be compared to that of other images. The two following approaches are used to represent a feature connected to a different image structure. The first uses hand-drawn elements in a single-stage network, and the second automatically learns features in green pixels through a multi-stage network.

Lily Guo talked about classifying natural images using a biologically stimulated model in [3]. The technique makes use of widely recognised, related advancements in the human brain's inference mechanisms and visual information processing. This approach is mostly used for natural categorisation and image analysis. This system is made up of three crucial parts, including a clustering of visual information unit, a unit for knowledge structure, and a unit for biologically inspired visual selective attention [4]. It automatically extracts significant associations between photos using low-level information in the images. In order to classify images more accurately, the system mimics the limitations of the human visual system.

CNN employs multi-layer convolution to build an engineering feature and integrate these features internally in traditional image recognition methods [5]. Both the integration layer and the fully integrated layer with SoftMax are utilised. Math can be done using Google's TensorFlow open source package, which focuses on machine learning math. The scientific community in the field of mechanical learning across the globe has paid greater attention to and shown more admiration for the recently released second version of Google's practical learning programme. TensorFlow provides several benefits, including excellent code, high adaptability, and accessibility, and the TensorFlow team on GitHub makes this startup job simple.

Three main categories of machine learning algorithms; supervised reading, supervised reading, and reinforced reading can be identified. We employ supervised reading in our model. With the use of mapping operations that map out the intended output, the dependent (target) will be predicted utilising a distinct independent (predictor) in the supervised reading process. A hybrid CNN-RNN model for video sentiment analysis was proposed [6]. The RNN was used to record the temporal dependencies while the CNN was utilised to extract spatiotemporal properties from videos. A three-stream CNN model for video sentiment analysis was proposed by [2]. The RGB frames, optical flow pictures, and motion boundary histograms were the three input streams employed by the model. On the Acted Facial Expression in the Wild (AFEW) database, the suggested model demonstrated cutting-edge performance.

A multi-scale temporal convolutional network (MSTCN) was suggested by [7] for the analysis of video sentiment. For the purpose of extracting features at various time scales, the MSTCN employed multiple temporal convolutional layers. The suggested model performed at the cutting edge on the AffectNet database. A spatial-temporal attention-based LSTM-CNN model for video sentiment analysis was put forth by Chen et al. in 2020. The most significant spatiotemporal elements in films were captured using the suggested model's spatial-temporal attention mechanisms. The suggested model performed at the cutting edge on the AffectNet database.

An enhanced attention method for video sentiment analysis was put forth by Wu et al. in 2021. The most significant spatiotemporal elements in films were captured using the proposed model's spatial and temporal attention mechanisms. The suggested model performed at the cutting edge on the EmoReact database. Images have the power to evoke complex emotions in viewers. Recent years have seen a significant increase in research efforts devoted to emotional image content analysis (AICA), due to the exponential rise of visual data. In this essay, we provide a thorough evaluation of contemporary approaches to two major problems: affective gap and perception subjectivity. It starts by providing an overview of the main emotion representation models that have been extensively used in AICA. A brief description of the existing datasets that are available for examination is provided. In the following section, they review and contrast the representative techniques to emotion feature extraction, customised emotion prediction, and emotion distribution learning.

On ideal data sets, innovative superior resolution algorithms have impressively performed despite blur and distortion. However, the majority of these techniques rely on simple bicubic downsampling to produce training low-resolution (LR) and high-resolution (HR) pairs from high-resolution pictures which may miss out on frequency-related details. So as a consequence, these techniques never succeed in resolving real-world images. Words and phrases used on social media reflect people's views on various products, solutions, authorities, and events. The goal of sentiment analysis, an aspect of natural language processing, is to recognise positive or negative qualities from social media posts. Because of the exponentially expanding needs of government organisations, corporate businesses, and individuals, researchers are motivated to finish their sentiment analysis study.

For the purpose of optimising sentiment analysis Naive Bayes, J48, BFTree, and OneR were four state-of-the-art machine learning classifiers used in this work. Three manually created datasets—two from Amazon and one from IMDB movie reviews—are used in the testing. The efficacy of these four classification techniques is examined and compared. Nave Bayes was shown to be quite quick at learning, even if OneR appears more promise in achieving the accuracy of 91.3% in precision, 97% in F-measure, and 92.34% in correctly classified instances.

Due to the complexity of the languages, sentiment analysis is one of the crucial and difficult issues in the field of artificial intelligence. Rule-based and machine learning-based models are more common. Existing models, however, have struggled to accurately categorise irony, sarcasm, and subjectivity in texts. In this study, we see them deploy cutting-edge sentiment analysis machine learning techniques to an open IMDB dataset and assess how well they work. There are numerous examples of irony and sarcasm in the dataset. Convolutional neural networks (CNN), bag of tricks (BoT), transformer-based models, and long-short term memory (LSTM) are constructed and assessed. Additionally, we looked at how hyper-parameters affected the models' accuracy.

A critical task in natural language processing is emotion recognition from text, which has huge implications for a variety of fields, including artificial intelligence, human-computer interaction, and others. Emotions are physiological ideas that people experience in response to experiences. Analysis of these emotions without consideration for voice and facial expressions is crucial and necessitates a supervisory technique for accurate emotion interpretation. Despite these obstacles, it's important to recognise human emotions as they increasingly express themselves through abusive text on social media platforms like Facebook, Twitter, etc. In this essay, they suggest categorising a large number of tweets according to their mood. Here, they classify an expression's feelings into positive or negative emotions using deep learning algorithms.

The types of negative emotions are: rage, boredom, emptiness, hate, sadness, and concern. Enthusiasm, fun, happiness, love, neutral, relief, and surprise are further subdivided into the positive emotions. They explored and evaluated the method employing long short-term memory and recurrent neural networks on three distinct datasets to show how to attain high emotion categorization accuracy. According to an in-depth study, the system enhances emotion prediction on the LSTM model with 88.47% accuracy for positive/negative classification and 89.13% and 91.3% accuracy for positive and negative subclass, accordingly.

We can learn useful information from the research of public opinion. There are many uses for sentiment analysis on social networks like Twitter and Facebook, which has developed into a potent tool for understanding user opinions. However, difficulties with natural language processing (NLP) are impeding sentiment analysis' effectiveness and accuracy. Deep learning models have been shown to be a promising answer to NLP's problems in recent years. The most recent research using deep learning to address issues with sentiment analysis, like sentiment polarity, is reviewed in this publication. Models constructed using term frequency-inverse document frequency (TF-IDF) and word embedding have been applied to a variety of datasets. The experimental results for various models and input features have also been compared in this work.

3 Methodology

Automated extraction, analysis, and comprehension of significant data from imagery is the focus of the interdisciplinary field of computer vision, which combines machine learning and artificial intelligence. With modern technological breakthroughs, digital content for videos and photos is booming. Understanding and processing images is a major challenge for computers in computer vision when compared to humans. So, using human assistance, the classification of images will be carried out. We utilised the 35887 48x48 pixel greyscale photos divided into 5 classes contained in the facial expression recognition dataset (Fig. 1).

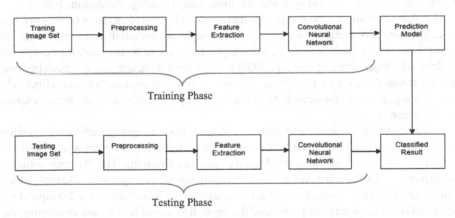

Fig. 1. Flow Diagram

The dataset will require us to make subtle differentiations since the images are somewhat closely related between specific classes. For example, some images under the 'sadness' class can be misconstrued as images under the 'disgust' class. A typical feedforward neural network would be unable to detect these subtle differentiations and would lead to the corresponding model having an unsatisfactory performance on this particular dataset. For example, to work on a standard photograph shot on a mobile phone, hundreds of millions of neurons would be required for a conventional feedforward network. We could scale the image down to incorporate the usage of a NN but that would lead to the image losing considerable substantial data. It is clear that traditional methods are of no consequential use. This is where the main methodology of this project comes into play; Convoluted Neural Networks (Fig. 2).

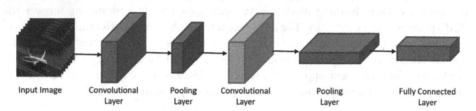

| Input Image | Convolutional Layer | Pooling Layer | Convolutional Layer | Pooling Layer | Fully Connected Layer |

Fig. 2. Typical CNN Architecture

OS, torch, torchvision are the libraries used along with a few utility functions like random_split, download_url etc. The dataset is split according to the sklearn split function. Coming to the preprocessing of the data, For loading the data as PyTorch tensors, we utilise the ImageFolder class of Torchvision.. A pytorch tensor is basically a generic numerical n-dimensional array to be used for arbitrary computation. The data or 'images' are read using PIL(Python Imaging Library).

Every element corresponds to a tuple that consists of a label and an image tensor. The label indicates which class the element belongs to. Each image tensor has the shape: (1, 48, 348). The data is then split into the three sets; Training, Validation, Testing.

We should examine how images are mathematically stored in terms of data prior to looking at how convolution processes operate. Images are essentially very large matrices of numbers, where each number denotes a pixel's brightness. In contrast to greyscale models, which only have one matrix, RGB models have three separate matrices because they represent three different colours. Because it is a compromise between effectively storing image data and the sensitivity of the human eye, the values within these matrices vary between 0 to 255.

Contrary to feedforward operations, we refrain from assigning each pixel a unique weight when performing convolution operations. However, we employ a something known as a kernel, which is essentially a 3x3 matrix of weights. The 2D convolution is a relatively straightforward process: as previously noted, we begin with a kernel, which is just a small matrix of weights. The kernel iteratively "slides" across the 2D input data, elementwise multiplying the portion of the input it is currently on, and combining the outcomes into one output pixel (Fig. 3).

The kernel replicates this procedure for each site it passes over, transforming one 2D feature matrix into the next 2D feature matrix and maintaining spatial features in the process. The input features are effectively positioned nearly in the same location as the output pixel on the input layer, and the output features are the weighted sums of those features, where the weights are values from the kernel itself. This is referred to as

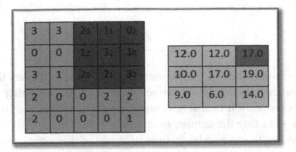

Fig. 3. Convolutional Operation

"feature extraction".

$$G[m, n] = (f * h)[m, n] = \sum\nolimits_j \sum\nolimits_k h[j, k] f[m - j, n - k]$$

By using the method above, where "f" stands for the input image and "h" for our kernel, we can calculate the feature maps. The rows and columns of the result matrix's indexes are denoted by the numbers m and n, respectively. The presence or absence of an input feature inside this nearly similar position is established by whether or not it is in the part of the kernel that produced the output. This implies that the number of input features are merged to create a new output feature depends on the size of the kernel. The convolution layers implement the following two methods: Strides and Padding:

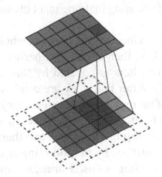

Fig. 4. Padding

The edges are basically cut off during the sliding operation, turning a 5x5 feature matrix into a 3x3 one. Since there is little in addition to these edges for the kernel expand to, the edges' pixels are never at the kernel's centre. This causes the output to differ in size from the input, which isn't optimal because we'd prefer for the output to be the same size as the input. The idea of padding is used to solve this issue. Padding effectively adds extra, bogus pixels to the edges (often with the value of 0). This enables the original edge pixels to slide with the kernel's centre, giving an output that is precisely the same size as the input.

$$p = \frac{\hat{f} - 1}{2}$$

The above formula, wherein p represents padding and f indicates filter dimension, must be true in order to maintain the former position. When a smaller output than the input is needed, convolution neural networks frequently reduce the size of the spatial dimensions through raising the number of channels. One approach to do this is to utilise a stride since step length can also be thought of as one of the convolutional layer hyperparameters. Stride essentially means bypassing part of the kernel's slide positions. To slide across one pixel apart, use a stride of 1. A two-pixel slide is equivalent to a stride of 2, and so on. Pooling can also be used to accomplish this.

$$n_{out} = [\frac{n_{in} + 2p - f}{s} + 1]$$

The method above can be used to determine the size of the output matrix, padding and stride included. The methods discussed above deal with only single channel inputs within the image tensor. However, we are dealing with multiple channels i.e. 3. As we go deeper into the convolutional network, the number of channels progressively increases. For an image tensor with 1 channel, the terms filter and kernel are interchangeable as they refer to the same Fig. 4 Padding concept in that particular context. In the general case, they are different entities. The term "filter" refers to a group of kernels, each of which is distinct and is in charge of every input channel. One filter will only yield one output in a convolution layer. A processed version of each input channel is created as each kernel passes through its corresponding input channel. Certain kernels might possess greater weights over others, Indicating that one input channel has a greater impact than the others.

A single output channel is produced by the filter as a whole since each of the processed versions of the input channels is then added together to constitute one channel. Any situation with any quantity of filters is covered by this. An output from a filter has its own distinct set of kernels. After that, they are combined to get the final result. In simpler words, if the usage of more than one filter on a single image is desired, then the convolution operation for each filter on the subsequent image is carried out separately. The results are then stacked on top of each other and then combined as a whole. The output tensor dimensions are satisfied by the following equation. nf- number of filters, nc - number of channels, p - padding, s - used stride, n - image size.

$$[n, n, n_c] * [f, f, n_c] = [\left[\frac{n + 2p - f}{s} + 1\right], \left[\frac{n + 2p - f}{s} + 1\right], n_f]$$

There are two steps in forward propagation within a layer of convolution. Computing the intermediate value— basically the outcome of convolutioning the input data from the preceding layers into a tensor (W) that contains the filters—and subsequently incorporating the bias (b) constitute the first phase. Z serves as the symbol for the intermediate value. The second phase in forward propagation involves applying a non-linear activation function (referred to by g) to Z, the intermediate value

$$z = W.A + b \qquad A = g(Z)$$

Calculating derivatives serves the purpose of later using them to adjust the parameter values in a procedure known as gradient descent. Backpropagation describes this situation. Calculations involve determining how modifications in the parameters will affect the feature maps that are produced and, ultimately, the outcome

$$dA = \frac{\delta L}{\delta A} dZ = \frac{\delta L}{\delta Z} dW = \frac{\delta L}{\delta w} db = \frac{\delta L}{\delta b}$$

dW and db are derivatives of parameters associated with the current layer with respect to the convolution operations. The intermediate value dZ is obtained by the input tensor having had a derivative of the activation function applied to it.

$$dZ = dA.g\prime(Z)$$

When dealing with backpropagation of the convolution itself, a matrix operation called full convolution is utilised. This operation is represented by the following, with W denoting the filter, and dZ[m,n] being a scalar belonging to a partial derivative from the previous layer. As we enter one convolution layer after another, the process keeps repeating. Along with convolutional layers, we additionally use a layer called max-pooling that gradually reduces the height and width of each convolutional layer's output tensors. Our pool is a maximum of 2×2. As we keep stacking kernels, each kernel keeps learning about their outputs and the information specifications contained by them gets richer and richer; this is basically how a convolutional neural network works. After the convolutional layers, we pass the output feature map through a function which flattens the output feature map into a vector. We do this as we want a vector of size five, each individual element containing a probability of belonging in each of the Facial Emotion Recognition classes.

4   Result and Discussions

Results were obtained by experimenting with the CNN algorithm. Here batch size is kept as a constant value with each change in the number of epochs.

To achieve best results, some of the following changes were made in the architecture of neural network for each experiment:

1) Number of Layers: Four Convolution layers were found to offer the optimum depth to the model while softmax and relu were the activation functions used for introducing non-linearity.
2) Number of epochs: With the increase in the number of epochs, it is observed that the accuracy also increases. However, a high number of epochs would also result in overfitting, thus bringing the overall efficacy of the model in actual situations down. Thus it was found that for 80 number of epochs model works optimally.
3) Filters: The number of filters was variably proportional to the actual accuracy (Tables 1 and 2).

Table 1. Results obtained from the experiments

Experiment No	Epochs	Validation accuracy	Validation loss	Loss	Accuracy
1	10	0.5998	0.9634	0.7714	0.6862
2	30	0.6878	0.9592	0.3557	0.8674
3	50	0.6937	1.1426	0.1968	0.9304
4	80	0.7193	0.8579	0.3363	0.8728

Table 2. Comparison with related work

Related works	Dataset	Algorithms	Results
Kumar, Kumar, & Sanyal, 2017 [8]	FERC-2013	CNN	approx 90%
Kulkarni, Bagal, 2015 [9]	FACES	Gabor, Long Gabor	82%, 87%
Shan, Guo, You, Lu &Bie, 2017 [10]	JAFFE, CK+	KNN	65.11, 77.27
Minaee, &Abdolrashidi, 2019 [11]	FERC-2013	Attentional CNN	70.02%
Proposed Methodology	FERC-2013	CNN	92%

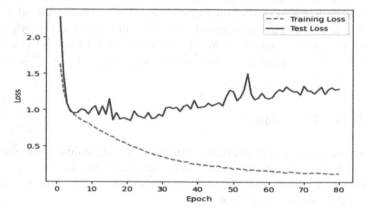

Plot 1. Loss vs Number of Epochs

It can be seen from the charts above that with each epoch, the accuracy improves and the loss reduces. Over the first several epochs, the training against testing curve for accuracy is optimal but it subsequently begins to stray from its optimum values (Plots. 1 and 2).

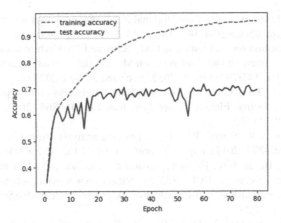

Plot 2. Accuracy vs Number of Epochs

5 Conclusion and Future Work

Sentiment analysis is a cutting-edge and mostly unexplored field with several real-world applications in the fields of business, science, and medicine. For businesses wanting to determine how the target audience would respond, this area has a lot of growth and implementation potential. Sentiment analysis can be used to automate a variety of processes, like handling restaurant reviews, among others. Neural networks are arguably the technological improvement with the maximum capability presently at the horizon [12]. By implementing neural networks, we could possibly handle almost any computational or statistical task more efficiently, and maybe one day with superior processing power than the human brain itself.

The implemented method was able to extract spatial and temporal data from video frames while gradually learning emotion patterns. The model developed by CNN based on a sizable video collection with a wide range of emotions and viewpoints was able to obtain an accuracy of 92%. CNNs offer a highly accurate, automatic, and reliable solution for emotion sentiment analysis that can be scaled to handle massive amounts of data in real-time. This makes it an effective tool for companies and organisations looking to understand client sentiments and enhance their offerings.

References

1. Lecun, Y., Bottou, L., Bengio, Y., Haffner, P.: Gradient-based learning applied to document recognition. Proc. IEEE **86**, 2278–2324 (1998). https://doi.org/10.1109/5.726791
2. Imani, H., Islam, M.B., Arica, N.: Three-Stream 3D deep CNN for no-Reference stereoscopic video quality assessment. Intell. Syst. Appl. **13**, 200059 (2022). https://doi.org/10.1016/j.iswa.2021.200059
3. Lily, G., Shifei, D.: Deep learning research progress (2015)
4. Wang, S., Liu, Y., Wang, X.: A deep learning approach for video sentiment analysis. In: Proceedings of the 2018 IEEE International Conference on Multimedia and Expo (ICME), pp. 1–6 (2018)

5. Visualizing and Understanding Convolutional Networks I SpringerLink. https://link.springer. com/chapter/https://doi.org/10.1007/978-3-319-10590-1_53. Accessed 19 Sept 2023

6. Video-based emotion recognition using CNN-RNN and C3D hybrid networks I Proceedings of the 18th ACM International Conference on Multimodal Interaction. https://dl.acm.org/doi/ https://doi.org/10.1145/2993148.2997632. Accessed 26 Sept 2023

7. Gan, Z., Zhou, J., Tang, G.: Temporal convolutional networks interval prediction model for wind speed forecasting. Electric Power Syst. Res. **191**, 106865 (2021). https://doi.org/10. 1016/j.epsr.2020.106865

8. Kumar, G., Kumar, R., Sanyal, P.G.: Facial emotion analysis using deep convolution neural network, pp. 369–374 (2017). https://doi.org/10.1109/CSPC.2017.8305872

9. Kulkarni, K.R., Bagal, S.B.: Facial expression recognition. 2015 International Conference on Information Processing (ICIP) (2015). https://www.semanticscholar.org/paper/Facial-expression-recognition-Kulkarni-Bagal/d4b4020e289c095ce2c2941685c6cd37667f5cc9. Accessed 26 Sept 2023

10. Ke, S., Jungi, G., Wenwan, Y., Di, L., Rongfan, B.: Automatic facial expression recognition based on a deep convolutional-neural-network structure. https://ieeexplore.ieee.org/abstract/ document/7965717/. Accessed 26 Sept 2023

11. Minaee, S., Minaei, M., Abdolrashidi, A.: Deep-emotion: facial expression recognition using attentional convolutional network. Sensors **21**(9), 3046 (2021). https://doi.org/10.3390/s21 093046

12. Uke, S., Zade, A.: An enhanced artificial neural network for hand gesture recognition using multi-modal features. Computer Methods in Biomechanics and Biomedical Engineering: Imaging & Visualization. https://doi.org/10.1080/21681163.2023.2227735

Deep Learning-Based Solution for Intrusion Detection in the Internet of Things

Akhil Chaurasia[1]([✉]), Alok Mishra[1], Udai Pratap Rao[2], and Alok Kumar[1]

[1] Sardar Vallabhbhai Institute of Technology, Surat, India
akhilchaurasia47@gmail.com
[2] National Institute of Technology, Patna, India

Abstract. Securing the Internet of Things-based environment is a top priority for consumers, businesses, and governments. There are billions of devices connecting and sharing data; an attack might cost billions of dollars. As a result, it's important to protect the IoT network from external and internal threats. There is no way to guarantee that all vulnerabilities will be fixed with a single solution or that no additional flaws will be discovered. This paper proposes a deep learning-based solution to detect network intrusion in an IoT network to better prepare for network attacks. The proposed solution achieves the optimal tradeoff between accuracy and model weightage and ensures it is well-suited for resource-constrained IoT devices. The proposed solution uses a reduced data set for training produced by incremental PCA with LSTM, GRU, and BiLSTM. The proposed solution reduced the training time significantly while retaining the accuracy of 98.17% with GRU, 98.12% with LSTM, and 98.23% with BiLSTM, and the results show that the proposed model has better performance in training the model for detecting network intrusion in an IoT network.

Keywords: Intrusion Detection · Deep Learning · IoT · IPCA

1 Introduction

The Internet of Things (IoT) is a network of interconnected physical devices and sensors that allows data to be shared over the Internet. IoT networks have proven to be particularly effective at collecting, analyzing, reporting, and forecasting data for use every day [18]. However, it offers several benefits to the community, but it also introduces some security vulnerabilities connected with IoT devices due to their worldwide connectivity, short battery life, ad-hoc nature, and mobility. Therefore, network security needs ongoing surveillance and analysis. Also, it's important to make the most of efforts to provide and maintain services, protect sensitive data, and be ready for any unexpected problems while monitoring and preventing cyber attacks on IoT networks.

A few network-based solutions are deployed to ensure security from attacks, such as Network Intrusion Detection Systems (NIDS) and firewalls [14]. For

R. Muthalagu et al. (Eds.): CINS 2023, CCIS 1978, pp. 75–89, 2024.
https://doi.org/10.1007/978-3-031-48984-6_7

securing IoT devices, firewalls are not feasible due to the high demand for computation power needed to analyse and filter the packets, which IoT devices cannot fulfill. NIDS monitors the traffic and alerts the administration in case of an intrusion into an IoT network. NIDS is placed in a strategically ideal position such that all the device's traffic monitoring can be done. NIDS can be classified as signature-based or anomaly-based NIDS, but these conventional approaches are infeasible because of the large number of network communication protocols among IoT nodes, and they can't detect zero-day attacks and needs continuous updating of rules for signature-based intrusion detection systems to be effective.

1.1   Motivation and Contribution

Machine learning requires a small dataset for training but provides less accuracy than deep learning, which can extract significant patterns from the dataset independently. The IoT industry finds Many security concerns unaddressed because of memory and computational capabilities. As a result, the objective is to apply a deep learning solution to investigate deep learning capabilities and develop a lightweight, adaptable model for IoT networks for intrusion detection. The significant contributions of this paper are as follows:

- Build a deep learning-based intrusion detection model for an IoT network.
- Reduce the training time of an IDS with Incremental Principal Component Analysis (IPCA).
- Achieve a tradeoff between binary classification model performance and accuracy improvements.

1.2   Organization

This paper is arranged into five sections. The basic introduction to the problem is given in the above section. The remainder of the paper is organized as follows. Section 2 contains the related work. Section 3 describes the preliminaries used in the proposed network model. Section 4 provides the proposed algorithm in detail. The result and observations are discussed in Sect. 5. At last, the conclusion is briefed.

2   Related Work

In recent years, there have been several security attack demonstrations against IoT networks, and researchers have done some work to tackle the problem. In this section, we will highlight some of the already existing work.

Krimmling et al. [9] concentrate on reducing security vulnerabilities associated with IoT network protocols. Their findings suggest that a hybrid approach to intrusion detection is preferable, combining rule-based and anomaly-based intrusion detection. Zhao et al. [21] proposed Propagation Neural Network

(BPNN) based model to determine whether or not the communication is malicious after the features have been chosen. They have a success rate of over 97%, although network traffic training takes nine weeks.

To construct a CNN-based IDS, Xiao et al. [19] used Batch Normalization with the KDD-99 dataset. The suggested framework used an auto-encoder (AE) network to reduce the number of dimensions. Riyaz et al. [15] used the KDD-99 dataset to create an IDS using a CNN architecture for wireless networks. A novel coefficient-based feature selection technique (CRF-LCFS) was used in the framework, which improved the model's detection accuracy and computation times.

Naseer et al. [12] proposed an intrusion detection model for anomaly detection in the IoT network. They implemented and trained three deep learning models: Deep Convolutional Neural Networks (DCNN), Long-Short Term Memory (LSTM), and Convolutional Autoencoders. They used the NSLKDD dataset for training. Their finding shows that deep learning is very efficient for anomaly detection.

Chawla et al. [16] concentrated on identifying anomalies in the IoT network to identify zero-day threats. They employed Deep Belief Network (DBN) for anomaly detection. A simulation dataset created with the COOJA network was used to train the model. Wang et al. [18] used state-of-the-art machine learning algorithms for IoT security. Their results indicate that SVM and K-means perform best on the network traffic analysis task.

Zhang et al. [20] presented an algorithm combining a genetic algorithm and a deep belief network for intrusion detection. Deep belief networks decide the number of neurons and layers to be used, and genetic algorithms create different network structures for different kinds of attacks. Hindy et al. [5] used an autoencoder to detect zero-day attacks. The result was tested using NSL-KDD and CICIDS 2017. The CICIDS2017 zero-day detection accuracy for DoS (Golden-Eye), DoS (Hulk), Port scanning, and DDoS attacks is 90.01%, 98.43%, 98.47%, and 99.67%, respectively.

Latif et al. [11] proposed a deep random neural (DRaNN) Gradient descent-based approach for the Industrial Internet of Things (IIoT), and they have claimed a low false-positive rate with an accuracy of 99.54% and detection rate of 99.41% over nine different kinds of attacks.

Parimala et al. [13] proposed a feature reduction technique with a combination of Conditional Random Field (CRF) and Spider Monkey Optimization (SMO), and CNN is used for classifying the traffic as an attack or normal. Imtiaz et al. [17] used LSTM, GRU, and BiLSTM for detecting the Intrusion detection they focused on Binary classification as well as Multiclass classification for detecting the attack kind their work on NSL-KDD and BoT-IoT dataset showed very high accuracy.

Conventional approaches such as signature-based and anomaly-based are infeasible because of the large number of network communication protocols among IoT nodes. This approach cannot detect zero-day attacks and needs continuous updating of rules for signature-based intrusion detection systems to be

Table 1. Recent related work on deep learning for intrusion detection

Article	Model	Dataset	Accuracy
[7]	ANN	CTU-13	99.94%
[10]	LSTM	KDD99	98.88%
[1]	LSTM Autoencoder	UNSW-NB15	98.00%
[14]	BiLSTM	KDD	99.70%
[2]	LSTM-GRU	BoT-IoT	99.76%
[2]	LSTM-GRU	NSL	99.14%
[4]	LSTM	CICIDS2017	99.55%
[3]	CNN	InSDN	97.50%
[6]	BiLSTM	NSLKDD	94.26%
[17]	LSTM	NSLKDD	99.91%
[17]	GRU	NSLKDD	99.91%
[17]	BiLSTM	NSLKDD	99.92%
[17]	LSTM	BoT-IoT	99.90%
[17]	GRU	BoT-IoT	99.93%
[17]	BiLSTM	BoT-IoT	99.96%

effective. An anomaly-based approach gives high false-positive cases because of the large number of variances in network communication patterns in an IoT network. An anomaly-based intrusion detection system can detect unusual attacks quickly but cannot identify the type of attack. However, any scenario that does not fit the regular pattern has been classified as an intrusion. Furthermore, investigating all aspects of normal behavior is not an easy task.

Machine learning approaches have given less accuracy compared with deep learning models. State-of-the-art models such as LSTM, GRU, BILSTM, CNN, and others have the researcher's interest and have given high accuracy after training. But the limiting factor for most of the papers is the model's deployable capacity on IoT devices. From the literature survey, we observed that most of the works talk about how accurate the system was after training and testing, but they didn't talk about how long training took and didn't compare memory and processing needs. Table 1 summarises the accuracy of some recent work using various state-of-the-art models using different datasets.

3   Proposed Model

This section will provide a method and a general approach that will lead us to the final solution to secure the IoT network from ID attacks and the dataset used to build the intelligent model. IoT devices can't run security software like computers can because they have low processing power and very little memory space.

The model's main goals are not to compromise security and to be lightweight, allowing it to be deployed on devices with low computational power.

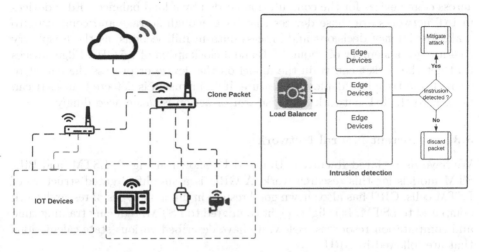

Fig. 1. Proposed Network Architecture

3.1 IDS-Dataset

For model training, we used the NSL-KDD and BoT-IoT datasets. The NSL-KDD data set is a new version of the KDD-99 data collection. This is a useful benchmark data set for comparing different intrusion detection methods proposed by different researchers. A feature description of the KDD dataset is described below.

- The KDD dataset described 24 attacks which are divided into four classes namely User to Root (U2R), Remote to Local (R2L), Denial of Service (DoS), and probe.
- A single sample as "normal" or with one of the attack types represented by 41 features.

3.2 Proposed Network Architecture

Many protocols exist for wireless communication protocols, such as ZigBee, Wi-Fi, Bluetooth, or using the CoAP protocol for resource-constrained devices in an IoT network. The flexible nature of the wireless protocol allows IoT devices and sensors to be placed anywhere within the protocol range. Fig 1. shows the network architecture where the intrusion detection model can be deployed. IoT sensors and devices generally communicate with the IoT hub for processing information or uploading data onto the cloud. The IoT hub can monitor incoming and outgoing packets in our IoT network. This hub can clone the packet and send the packet to the intrusion detection agents to analyze the packet data and report an intrusion alert. Intrusion detection agents form a small network of edge devices and load balancers. The packets received are evenly distributed across edge devices for the computation needed by a load balancer. Edge devices in IoT networks are those devices that have enough storage and computing to make low-latency decisions and process data in milliseconds, as the Raspberry Pi 4 has got a memory of about 4 GB and a clock speed of 1.5 GHz. Edge devices that get the packet can train the model on the device or process the packet to extract and train and deploy the feature. If an intrusion is detected, an alert can be sent to the IoT network administrator or action is taken accordingly.

3.3   Recurrent Neural Network

We have used Gated Recurrent Unit (GRU) shown in Fig. 2, LSTM, and BiL-STM models for this research work. A GRU is a simplified gated structure of LSTM cells. GRU has also shown good results in different types of research work compared to LSTM. It is lightweight compared to LSTM, reducing training time and computation resources. Below, we have described various state calculations that are followed in GRU.

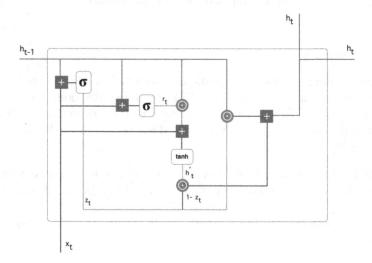

Fig. 2. Gated Recurrent Unit [8]

Candidate State

$$h'_t = g(W_{fh}x_t + W_{rh}(h_{t-1} \odot r_t) + \varphi_h) \tag{1}$$

It does the Hadamard product between the reset gate and the previous state output. This operation will determine what to remove from the previous state. It is then summed with the product of weight and input vectors.

Reset Gate

$$r_t = f(W_{fr}x_t + W_{rr}(h_{t-1}) + \varphi_r) \qquad (2)$$

The computation of the reset gate is similar to that of the update gate. Its main purpose is to decide how much past information to forget.

Update Gate

$$z_t = f(W_{fz}x_t + W_{rz}(h_{t-1}) + \varphi_z) \qquad (3)$$

x_t is given as an input to the network unit. It gets multiplied by its own weight W_{fz} same goes with h_{t-1}, information from the previous unit.

Current Gate

$$h_t = (1 - z_t) \odot \tilde{h}_t + z_t \odot h_{t-1} \qquad (4)$$

The final state of the current network unit is determined by the current memory state ht and the previous state h_{t-1}.

f(.) indicates sigmoid function and g(.) indicates tanh function below shows the curve and equation

Sigmoid Activation Function. The sigmoid function graph is shown as an S-shaped curve in Fig. 3. Given that its domain is the set of all real numbers and its range is between (0, 1). This function is also known as the squashing function. The sigmoid activation function is calculated as follows.

$$\frac{1}{1 + e^{-x}} \qquad (5)$$

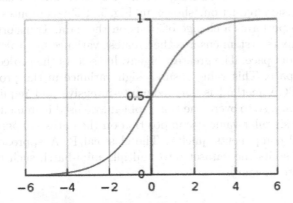

Fig. 3. Sigmoid Activation Function

Tanh Activation Function. Figure 4. shows the Tanh activation, also called the hyperbolic tangent activation function. The function takes a real number as input, and the output value ranges from −1 to 1.

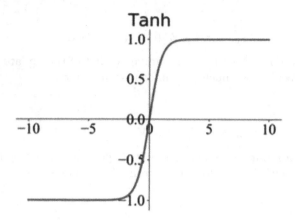

Fig. 4. Tanh Activation Function

The figure shows the S shape of the activation function where a higher positive input will give an output close to 1, and with a negative input, the resultant output is closer to -1. Below tanh shows the equation is shown.

$$\sigma(z) = \frac{e^z - e^{-z}}{e^z + e^{-z}} \tag{6}$$

4   Proposed Algorithm

To effectively train our model in the IoT network, it is crucial to keep account of the number of neurons in each layer. The proposed solution uses incremental PCA for dimension reduction, shown in Fig. 5. PCA computes the principal component and performs a change of basis on the data. Transforming the data to a new coordinate system ensures the greatest variance by projecting the data into a new vector space. The greatest eigenvalue selects the projected data onto a new vector space. This value ensures high variance in the projected vector. However, the PCA method is very compute-intensive and requires the whole dataset in memory, so to overcome this problem, we used incremental PCA. This randomizes the singular value decomposition on the dataset, approximating the first k principal components quickly. The classical PCA approach, along with randomization, splits the dataset into multiple mini-batch such that one mini-batch can fit into memory.

Algorithm 1: Intrusion detection

Input: Train and Test data set
Output: Intrrusion Detection model

1 Define $dataset^{train}, dataset^{test}$
2 **while** *rows in $dataset^{train}, dataset^{test}$* **do**
3 $\quad$ $X^{train}, X^{test} \leftarrow$ select-numerical -feature
4 $\quad$ $\bar{X}^{train}, \bar{X}^{test} \leftarrow$ select-non-numerical -feature

5 $X^{train}, X^{test} \leftarrow Normalize(X^{train}, X^{test})$
6 $\bar{X} \cdot \bar{X}^{train}, \bar{X}^{test} \leftarrow OneHotEncoding(\bar{X}^{train}, \bar{X}^{test})$
7 $X^{train} \leftarrow append(X^{train}, \bar{X}^{train})$
8 $X^{test} \leftarrow append(X^{test}, \bar{X}^{test})$
9 $X^{train} \leftarrow Incremental - PCA - Dimension - Reduction(X^{train})$
10 $X^{test} \leftarrow Incremental - PCA - Dimension - Reduction(X^{test})$
11 Initialize Sequential GRU/LSTM/BILSTM deep Learning Model
12 compile binary creossentropy clustering
13 m $\leftarrow$ sequential-deep-learning -model
14 Training $\leftarrow m \in X^{train}$
15 **if** *training completed* **then**
16 $\quad$ predict $\leftarrow m \in X^{test}$
17 $\quad$ **if** *Predictions are correct* **then**
18 $\quad\quad$ deploy model
19 $\quad$ **else if** *Pediction are incorrect* **then**
20 $\quad\quad$ Retrain model

Using incremental PCA avoids the high demand for memory and computing power needed in the case of classical PCA.

The 1,25,973 samples are used for training with a validation split ratio of 0.2 and 22,544 samples for testing, and subsequently, this dataset is split based on numerical features and non-numerical features. Steps 2–4 show the split. A normalized numerical dataset, also known as Min-Max scaling, reduces the feature to a range of 0 to 1, and OneHotEncoding is applied to non-numerical features. The final train feature set is obtained by appending the normalized data and one hot encoded feature. The steps are done in 4–7. The high-dimensional feature set is reduced to a lower-dimensional feature set by applying incremental principal component analysis and selecting the top features from the reduced set, giving high variance in steps 8–9. Step 10–12 The model is initialized with layers described in Fig. 3.5 and trained with the training dataset. The model's accuracy, precision, F1 score, and recall are measured on the test dataset. If the results are good enough, the model is deployed, but if not, steps 13 through 17 show how to retrain it.

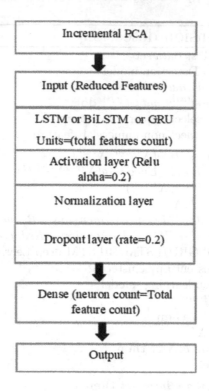

Fig. 5. Enhanced Flow of Proposed Approach

5   Performance Analysis and Results Discussion

5.1   Evaluation Metrics

Our analysis led us to conclude that employing GRU models is justified for resource-constrained IoT devices, as they effectively reduce feature dimensions so a trade-off can be achieved between accuracy and speed, leaving the IoT devices less vulnerable to intrusions. We have compared GRU with LSTM and BIL-STM models for performance evaluation. The results favored GRU in terms of evaluation matrix and training time, which are described in Table 2 and Table 3. Accuracy, precision, recall, and F1 score validate the suggested LSTM, BiLSTM, or GRU models. The results are obtained by implementing them in Keras and training them on the Intel Core i3-6100U CPU @ 2.30GHz, 4.0GB RAM, and Intel HD Graphics 520. We have used the below-listed formulas to calculate the accuracy, precision, recall, and F1 score. The training and validation accuracy is shown in Fig. 6. where the training and validation loss is shown in. Figure 7.

$$Accuracy = \frac{(TP + TN)}{(TP + FP + TN + FN)} \tag{7}$$

$$Precision = \frac{(TP)}{(TP + FP)} \qquad (8)$$

$$Recall = Sensitivity = \frac{(TP)}{(TP + FN)} \qquad (9)$$

$$F_1 Score = 2X\frac{(Precision X Recall)}{(Precision + Recall)} \qquad (10)$$

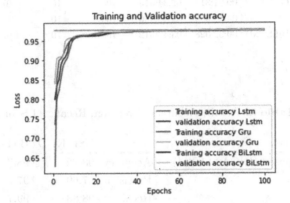

Fig. 6. Training and Validation Accuracy

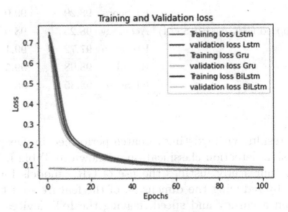

Fig. 7. Training and Validation Loss

5.2   Result Discussion

Table 2. Model Train Time Comparison

Model	Train Time (min:sec:ms)		Epochs		Final Train Features	
	NSL-KDD	BoT-IoT	NSL-KDD	BoT-IoT	NSL-KDD	BoT-IoT
LSTM [21]	02:24:233	03:09:219	100	10	93	10
Proposed (IPCA+LSTM)	00:22:364	02:36:913	100	10	10	5
GRU [21]	02:20:274	03:00:586	100	10	93	10
Proposed (IPCA+GRU)	00:21:318	02:34:913	100	10	10	5
BiLSTM [21]	04:18:566	04:42:990	100	10	93	10
Proposed (IPCA+BiLSTM)	00:32:028	03:48:312	100	10	10	5

Table 3. Result: Accuracy, Precision, Recall, F1 Score

Dataset		NSL-KDD	BoT-IoT
Proposed (IPCA+LSTM)	Accuracy	98.12	98.7
	Precision	97.69	97.5
	Recall	98.83	99.7
	F1 Score	98.25	98.9
Proposed (IPCA+GRU)	Accuracy	98.17	99.3
	Precision	97.79	98.1
	Recall	98.78	99.5
	F1 Score	98.29	99.0
Proposed (IPCA+BiLSTM)	Accuracy	98.23	98.9
	Precision	97.72	99.1
	Recall	98.98	99.5
	F1 Score	98.35	98.7

Comparing the results with existing research performed by Deep learning algorithms on intrusion detection classification as shown in Table I. We came to an observation that our study justifies the use of GRU models for light computing IoT devices by reducing the dimension of the feature so a trade-off can be achieved between accuracy and speed, leaving the IoT devices less vulnerable to intrusions. We have compared GRU with LSTM and BILSTM models for performance evaluation results favored GRU in terms of evaluation matrix and training time described in Table 2 and Table 3.

Figure 8. shows the PCA vs. IPCA memory requirement for feature reduction on the NSL-Kdd dataset, showing PCA needs more memory for feature

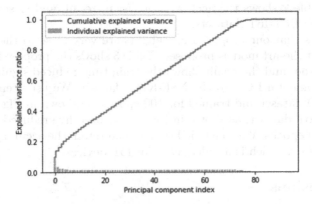

Fig. 8. Cumulative variance on NSL-KDD dataset

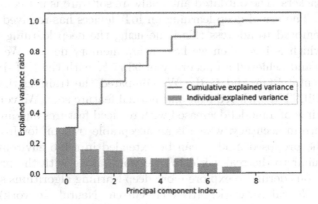

Fig. 9. Cumulative variance on Bot-Iot dataset

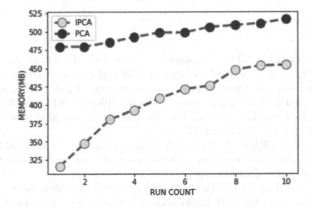

Fig. 10. Feature Extraction Memory Requirement PCA vs IPCA

reduction. Table 2 shows the accuracy, precision, recall, and f1 scores over the NSL-KDD and BoT-IoT datasets.

The results that our proposed model gets are very close to the results that earlier state-of-the-art models proposed. Table 3 shows the proposed model train time comparison, and the result shows the train time reduced significantly after applying incremental PCA to the NSL-KDD dataset. We used ten features for the NSL-KDD dataset and trained for 100 epochs, as shown in Fig. 9. Similarly, for the BoT-IoT dataset, as shown in Fig. 10, we used five reduced features and trained for ten epochs. We concluded that we can retain the accuracy and reduce the training time, which is an ideal case for IoT devices.

6   Conclusions

Due to their limited processing power and memory, IoT devices have frequently been spotted purposely compromising security. This makes them an attractive target for attackers. The outdated anti-malware software is not compatible with these devices. The use of deep learning on IoT devices has received some attention. We attempted to address this issue using the deep learning GRU model and IPCA, which is idle when we have a low-memory device. We reduced the training time and achieved an accuracy of 98.17 % with the NSL-KDD dataset using incremental PCA and GRU. We compared the training time for GRU, LSTM, and BiLSTM with reduced and normal feature sets. We concluded that the training time of a model decreased with reduced features and an overall very small difference in accuracy, which is an acceptable tradeoff for IoT devices. In the future, the proposed models can be extended in a lab environment, which should be similar to the real-world IoT network, along with the proposed approach we are considering to explore more deep learning algorithms such as CNN (Convolution Neural Network), RNN (Recurrent Neural Network), and Reinforcement learning combined with Incremental PCA, and finally, evaluate the performance with the real-world IoT attacks on a volatile node and measure the real-world effectiveness of the deep learning model.

References

1. Ashraf, J., Bakhshi, A.D., Moustafa, N., Khurshid, H., Javed, A., Beheshti, A.: Novel deep learning-enabled lstm autoencoder architecture for discovering anomalous events from intelligent transportation systems. IEEE Trans. Intell. Transp. Syst. **22**(7), 4507–4518 (2021). https://doi.org/10.1109/TITS.2020.3017882
2. Biswas, R., Roy, S.: Botnet traffic identification using neural networks. Multimedia Tools Appli. **80**(16), 24147–24171 (2021)
3. ElSayed, M.S., Le-Khac, N.A., Albahar, M.A., Jurcut, A.: A novel hybrid model for intrusion detection systems in sdns based on cnn and a new regularization technique. J. Netw. Comput. Appl. **191**, 103160 (2021)
4. Hai, T.H., Nam, L.H.: A practical comparison of deep learning methods for network intrusion detection. In: 2021 International Conference on Electrical, Communication, and Computer Engineering (ICECCE), pp. 1–6 (2021). https://doi.org/10.1109/ICECCE52056.2021.9514161

5. Hindy, H., Atkinson, R., Tachtatzis, C., Colin, J.N., Bayne, E., Bellekens, X.: Utilising deep learning techniques for effective zero-day attack detection. Electronics **9**(10), 1684 (2020)
6. Imrana, Y., Xiang, Y., Ali, L., Abdul-Rauf, Z.: A bidirectional lstm deep learning approach for intrusion detection. Expert Syst. Appl. **185**, 115524 (2021)
7. Joshi, C., Ranjan, R.K., Bharti, V.: A fuzzy logic based feature engineering approach for botnet detection using ann. J. King Saud University-Comput. Informat. Sci. (2021)
8. Kostadinov, S.: Understanding gru network. https://towardsdatascience.com/understanding-gru-networks-2ef37df6c9be (Accessed 24 June 2022)
9. Krimmling, J., Peter, S.: Integration and evaluation of intrusion detection for coap in smart city applications. In: 2014 IEEE Conference on Communications and Network Security, pp. 73–78. IEEE (2014)
10. Laghrissi, F., Douzi, S., Douzi, K., Hssina, B.: Intrusion detection systems using long short-term memory (lstm). J. Big Data **8**(1), 1–16 (2021)
11. Latif, S., Idrees, Z., Zou, Z., Ahmad, J.: Drann: a deep random neural network model for intrusion detection in industrial iot. In: 2020 International Conference on UK-China Emerging Technologies (UCET), pp. 1–4 (2020). https://doi.org/10.1109/UCET51115.2020.9205361
12. Naseer, S., et al.: Enhanced network anomaly detection based on deep neural networks. IEEE Access **6**, 48231–48246 (2018)
13. Parimala, G., Kayalvizhi, R.: An effective intrusion detection system for securing iot using feature selection and deep learning. In: 2021 International Conference on Computer Communication and Informatics (ICCCI), pp. 1–4 (2021). https://doi.org/10.1109/ICCCI50826.2021.9402562
14. Pooja, T., Shrinivasacharya, P.: Evaluating neural networks using bi-directional lstm for network ids (intrusion detection systems) in cyber security. Global Trans. Proc. **2**(2), 448–454 (2021)
15. Riyaz, B., Ganapathy, S.: A deep learning approach for effective intrusion detection in wireless networks using cnn. Soft. Comput. **24**(22), 17265–17278 (2020)
16. Thamilarasu, G., Chawla, S.: Towards deep-learning-driven intrusion detection for the internet of things. Sensors **19**(9), 1977 (2019)
17. Ullah, I., Mahmoud, Q.H.: Design and development of rnn anomaly detection model for iot networks. IEEE Access **10**, 62722–62750 (2022). https://doi.org/10.1109/ACCESS.2022.3176317
18. Wang, H., Barriga, L., Vahidi, A., Raza, S.: Machine learning for security at the iot edge - a feasibility study. In: 2019 IEEE 16th International Conference on Mobile Ad Hoc and Sensor Systems Workshops (MASSW), pp. 7–12 (2019). https://doi.org/10.1109/MASSW.2019.00009
19. Xiao, Y., Xing, C., Zhang, T., Zhao, Z.: An intrusion detection model based on feature reduction and convolutional neural networks. IEEE Access **7**, 42210–42219 (2019). https://doi.org/10.1109/ACCESS.2019.2904620
20. Zhang, Y., Li, P., Wang, X.: Intrusion detection for iot based on improved genetic algorithm and deep belief network. IEEE Access **7**, 31711–31722 (2019). https://doi.org/10.1109/ACCESS.2019.2903723
21. Zhao, J.W., et al.: Method of choosing optimal features used to intrusion detection system in coal mine disaster warning internet of things based on immunity algorithm. Vet. Clin. Pathol: A Case-Based Approach, 157 (2015)

Plant Protein Classification Using K-mer Encoding

K. Veningston[1][(⊠)], P. V. Venkateswara Rao[2], M. Pravallika Devi[1],
S. Pranitha Reddy[1], and M. Ronalda[1]

[1] Department of Computer Science and Engineering, National Institute of Technology Srinagar,
190006, Srinagar, Jammu and Kashmir, India
veningstonk@gmail.com
[2] Department of Computer Science and Engineering, GITAM School of Technology,
Visakhapatnam, Andhra Pradesh 530045, India

Abstract. Proteins play an important role in the human body and in plants. A lack of expertise in protein labeling in plants can make it extremely difficult to characterize and comprehend the precise roles and activities of different proteins. Furthermore, it restricts development in fields like biotechnology, disease resistance, and crop enhancement. The presented project focuses on plant protein classification, aiming to overcome the challenges arising from limited protein labeling knowledge. Advanced machine learning techniques, including various classification algorithms such as Logistic Regression, Decision Tree, K-nearest neighbors (KNN), Support Vector Machines (SVM), Random Forest (RF), Multinomial Naive Bayes (NB), AdaBoost, and XGBoost, are employed to accurately classify protein sequences into their respective families. This classification approach provides valuable insights into the functions and roles of proteins within plants, ultimately advancing our understanding of plant biology. This attempt offers new possibilities for advancement in critical sectors such as agriculture, drug discovery, and genomic research by eliminating the limitations associated with limited protein labeling knowledge.

Keywords: Genome Annotation · Peach (Prunus persica) Genome · Machine Learning · Protein Classification · K-mer Encoding

1 Introduction

Proteins are macromolecules that play crucial roles in the body. They are used to structure the body's organs and help in regulation and body functioning. They are composed of hundreds of amino acids that are covalently connected to each other. There are amino acids of 20 types, and the sequence predicts the protein's 3D structure and corresponding function. Combining four nucleotide bases ('A', 'T', 'G', and 'C') results in coded amino acids. Proteins play a crucial role in plant growth and development. They provide a variety of functions, including photosynthesis, biosynthesis, transportation, immunology, etc. A diverse range of proteins, including enzymes, structural proteins, and storage proteins, are produced by plants. *Enzymes* act as catalysts and are crucial for plant metabolism, *Structural Proteins* provides support and shape to plant cells, and *Storage proteins* stores amino acids for later use. Deoxyribonucleic acid (DNA) is a macromolecule composed of two polynucleotide chains that form a double helix structure by coiling around each other as shown in Fig. 1. It carries the genetic code, which is essential for all living things, to develop, function, grow, and reproduce. The two DNA strands are made up of nucleotides further stated to as polynucleotides. Each nucleotide is made up of a phosphate group, a deoxyribose sugar, and one of the four nitrogen-containing nucleobases (A, T, G, and C). Since more than 98% of human DNA is non-coding, it cannot serve as a guide for the construction of protein sequences. Due to the fact that they move in opposing directions, the two DNA strands are antiparallel. Transcription is the process of converting DNA nucleotides into ribonucleic acid (RNA) strands, with the exception of thymine (T), for which RNA substitutes uracil (U). Translation is the process by which these RNA strands use genetic instructions to direct the arrangement of amino acids in proteins.

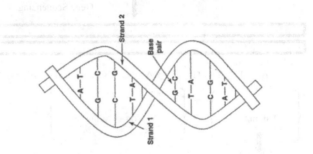

Fig. 1. DNA Double Helix Structure

1.1 Gene Expression

The genetic information of a cell is stored in DNA or RNA chemical form. A *Gene* is a portion or a segment of a DNA molecule and the *Genome* is the entire DNA in a cell. Each cell has a complex and closely controlled process from gene to protein. Transcription and translation are the two main phases. Transcription and translation work together to produce gene expression. The information contained in a gene's DNA is transferred to a similar molecule called RNA in the cell nucleus during the transcription process. Since messenger RNA (mRNA) transmits information from the nucleus to the cytoplasm, it is this type of RNA that contains the instructions needed to make a protein. The cell wall is where translation, i.e., the second step in turning a gene into a protein, takes place. The ribosome and the mRNA interact, and the ribosome "reads" the nucleotide sequence of the mRNA. Three nucleotide sequences make up a codon, which typically codes for one amino acid. One amino acid at a time, transfer RNA (tRNA), a kind of RNA, that assembles the protein. Until the ribosome encounters a "stop" codon (a three-nucleotide sequence that does not code for an amino acid), protein production continues. The transmission of information from DNA through RNA to proteins is one of the fundamental ideas in molecular biology ("central dogma").

1.2 Genome Annotation

Giving the sequence of the genome is never enough we need to know which part of the genome is coding and which part is non-coding as shown in Fig. 2. Here comes Genome Annotation, which is the process of describing the structure and function of a genome. Gene sequencing is the process of determining the nucleic acid sequence, or the arrangement of the nucleotides in DNA. Annotating a genome consists of three major steps; (*1*) Identifying the genome regions that do not contain protein-coding genes, (*2*) Gene Prediction - Process of identifying genome elements, and (*3*) Linking biological information to the elements.

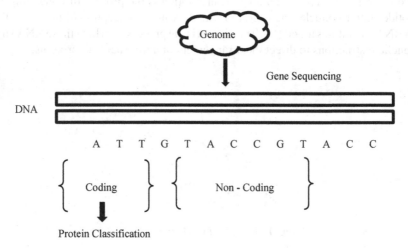

Fig. 2. Genome Annotation

1.3 Contributions

- The classification problem addressed in this paper helps gain insights into the functions and roles of different proteins within plants (specifically the *peach* plant), which has implications in agricultural genomic research.
- It demonstrates the critical role of proteins in *peach* plants and the challenges stemming from limited protein labeling knowledge. By focusing on plant protein classification, it aims to bridge this gap, which is important for understanding plant biology and precision agriculture.
- K-mer encoding represents plant genomic sequences by breaking them into fixed-length subsequences (K-mers), providing a condensed numerical representation for computational analysis of plant genomes.
- By circumventing the constraints of limited protein labeling knowledge, this paper enables informed decisions in agriculture in terms of disease resistance, crop enhancement, and allied fields dependent on a deep understanding of plant proteins.

2   Related Works

There have been several works related to plant protein classification that explore different approaches and techniques. These are a few examples of works related to plant protein classification. The field is evolving, and the efficiency of classification algorithms has been improving for identifying plant proteins.

2.1   Protein Classification Using Genome Neighborhood

Genomic neighborhood information is employed to find photosynthetic proteins and take the UniProt Knowledge Base/Swiss-Prot database [1]. Genome neighborhood network (GeNN) outperformed Random Forest and decision tree methods with an accuracy of 87%. As evidence of the model's potential to increase photosynthetic efficiency, it also showed its capacity to recognize novel photosynthetic proteins. The key outcome of this is the functional relationship between the neighbors of photosynthetic genes.

2.2   Classification of Plant Transcription Factor Proteins

Science still faces difficulties in categorizing amino acid sequences and comprehending the connection between amino acids and protein synthesis. To categorize the amino acid sequences of unidentified species, this study exploits the plant transcription factor database [1]. The model successfully classified transcription factor proteins in the kingdom of plants with a high success rate of 98.23% throughout tests. The hybrid model performs better than traditional long short-term memory – Convolutional Neural Network (LSTM – CNN)-based models due to its lightweight layers and shorter training time. The suggested model is improved by the usage of Word2Vec vectors.

2.3   Plant Allergenic Protein Prediction Based on Sequence

This paper comprehends the allergenic properties of dietary proteins. By combining supervised and unsupervised machine learning approaches, it is possible to predict if plant proteins may cause allergic reactions [2]. The method comprises rating descriptors and evaluating the effectiveness of their categorization. An SVM is utilized for partitioning, while a k-nearest neighbor (KNN) classifier is used for classification. For variable selection and final classification, the cross-validation (CV = 5) method is used to validate the KNN classifier. In order to address food allergies, the study emphasizes the necessity for a reliable and practical protein classification system.

2.4   Evaluating Plant Gene Models

True gene is an ML approach developed for gene model classification and reduces false positive annotations in gene prediction [3]. The Legume Information System (LIS) provided the Pisum annotated gene and protein dataset. The National Center for Biotechnology Information- Non-Redundant (NCBI-NR) database was compared to the annotated genes. It makes use of 41 protein-based traits, including amino acid and nucleotide sequences, and 14 genes. The Pisum genome was used for eXtreme Gradient Boosting (XGBoost) model training, which resulted in optimized models with 87–90% prediction accuracy and F-1 scores of 0.91–0.94.

2.5 Plant Vacuole Protein Prediction Based on Sequence

Prediction of subcellular localization is essential for comprehending gene functions in proteomes. This study addresses this issue by developing various compositions and position-specific scoring matrix (PSSM) based models, resulting in improved accuracy compared to previous methods [4]. They used the UniProt Knowledge Base/SwissProt database. The best model achieved approximately 63% accuracy on a blind dataset, surpassing current tools. To make the models accessible, they developed 'VacPred' [GUI-based software], compatible with Windows and Linux platforms. They reported an accuracy of 86.49%/87.84% and a sensitivity of 90.54%/93.24%.

2.6 DNA Sequence Classification using K-mer Counting

DNA sequences need to be classified in genomic research to determine the applicability of a new protein. This study utilizes machine learning algorithms to identify classes of DNA sequences based on nucleotide sequences. The open DNA sequence dataset was used to obtain the gene sequence dataset. Different datasets representing gene families are analyzed using substrings of defined lengths (determined by the k value) to capture sequence patterns. When they took the human dataset $k = 6$ they obtained the best accuracy 98.4%, and Precision 98.4%. The researchers obtained a gene sequence dataset about chimpanzees, dogs, and humans. The classification of DNA sequences within the framework of the analysis of biomedical data using deep learning (DL) has the ability to extract pertinent features from the input data [12]. They specifically used two different architectures namely CNN-LSTM, and CNN-Bi LSTM (Bidirectional LSTM). Both Label encoding and K-mer encoding strategies [5, 6, 8] were used for representing the DNA sequences. When the models were tested using a variety of classification criteria, the CNN and CNN-Bi LSTM with K-mer encoding both demonstrated good accuracy, scoring 93.16% and 93.13% on the testing data.

2.7 Long Terminal Repeats (LTR) Retrotransposons Classification Using K-mer Method

Although LTR retrotransposons are prevalent repeating sequences in plant genomes, the classification often involves laborious, manual procedures [7]. To overcome the problem, the researchers created a technique for classifying LTR retrotransposons and mapping them into certain families using K-mer-based ML algorithms. They used InpactorDB, which contains 67,241 LTR retrotransposon sequences from 195 plant species that have been systematically categorized into families. This dataset included sequences from Repbase, RepetDB, and Plant Genome and Systems Biology (PGSB). Their approach achieved an impressive 95% F1-Score.

2.8 Plant Protein Classification using Ensemble Classifiers

The method for forecasting the subcellular localization using several classifiers is enhanced in this research. The authors suggested an ensemble ML approach based on average voting [9, 13]. The dataset for testing and training was obtained from Plant-mSubP. They gather different features appropriate for each sort of localization, use feature selection to lessen dimensionality, and then train three different models. According to experimental findings, using the testing dataset, the suggested ensemble technique could correctly classify objects in 11 compartments with an accuracy of 84.58%.

3 Proposed Model

Plant protein classification using K-mer encoding involves representing protein sequences as fixed-length feature vectors based on the occurrence frequencies of subsequences called K-mers.

3.1 Problem Statement

Build a robust classification model that correctly maps amino acid sequences to protein families within the PlantGDB Database [11]. In this study, several models were used to solve the above problem. The model analyzes the input sequence data and predicts the protein class it belongs to with high accuracy and precision utilizing Machine Learning algorithms. By proposing an efficient and automated approach for protein family classification, this work aims to improve knowledge of protein structure and function.

3.2 Dataset Description

The PlantGDB is a data repository containing different plant species' genome sequences. In addition to sequence data, it also provides alignments and annotations [10]. *'Prunus persica [peach] genome'* comes with a *peptide file* (Ppersica_139_peptide.fa.gz) and the *annotation file* (Ppersica_139_annotation_info.text.gz). Table 1 shows the dataset statistics.

Table 1. Dataset statistics

Data characteristics	Count
# Samples	3824
# Features (length of the sequence)	1577562
# Classes	34

3.3 Dataset Preparation

The Peptide file considered for preparing data is in FASTA format (text-based) and it needs to be changed to CSV format for applying Machine Learning algorithms. For reading the FASTA file SeqIO.parse() function from *BioPython* Library is used. Extracted required attributes, stored them in a pandas data frame, and finally saved them in a file named 'peptide_data.csv' as shown in Fig. 3.

```
 1  >ppa010779m|PACid:17640292
 2  MAASNSGTVSLKLLIDTTRNKVLFAEAGKDFVDFLFTLLSLPAGTIIRLLSKDAMVGSLGKLYESVETLNDEYLQPNLNK
 3  DTLLKPKEPVAAGPNLLGLLTDVKSDAPKTIYRCSNGSCIYRISYVADDPKAICPGCKHSMTTAVTYVASPTSEVQATYS
 4  GAGKVGYVKGLVTYMIMDDLEVKPLSTISCITLLSRFSVKDVGVLEEKVVDLGMDEGVKLLKASLQTKSVLTQVFLR*
 5  >ppa004704m|PACid:17640293
 6  MSSHLYGYGATQSAAAAAAATAGLSSVYTSRTLTDPTLRYLSGSDPFASATDHHSRSSSMYLATSHLMSQSSWPAPDVEP
 7  GVPGVKRPSEALYHQSFMGAYNTIGQEAWYSALAKRPRYESASNLPIYPQRPGEKDCAHYMLTRTCKFGELCKFDHPIWV
 8  PEGGIPDWKEVPLVAPSESLPERPGEPDCPYFIKTQRCKFGMRCKFNHPKEKLAAAVASENADVFALPERPSEPPCAFYM
 9  KTGQCKFGATCKFHHPKDIQIPSAEQENKIGETGTTIQPEGTGFAVKLPVSFSPALLYNSKELPVRPGEPDCPFYLKTGS
10  CKYGATCRYNHPDRYAINPPIGAISHPIVAPPAAGLNIGVINPAASIYQTLAQPTVGGGQTVYPQRFGQIECDYYMKTGE
11  CRFGEQCKYHHPIDRSAVTLSTTKPVQQQNVKLTLAGLPRREGVAICVYYLKTGTCKYGATCKFDHPPPGEVMGMAASQG
12  ASGGEANDFTHEQQQ*
```

↓

	ID	PACId	Sequence
0	ppa010779m	17640292	MAASNSGTVSLKLLIDTTRNKVLFAEAGKDFVDFLFTLLSLPAGTI...
1	ppa004704m	17640293	MSSHLYGYGATQSAAAAAAATAGLSSVYTSRTLTDPTLRYLSGSDP...
2	ppa025459m	17640294	MDQQYSRDFPILSYILSRLDPESNPPLSPQLQETLLTQLPHLNHPK...
3	ppa025141m	17640295	MAMKQPHVIIFPFPLQGHMKPLLCLAELLCHAGLHVTYVNTHHNHQ...
4	ppa001905m	17640296	MCSSTNVPRWTPSPSPTRSLLASAVEGPSSKESHVLDDDHAEMKRQ...

Fig. 3. Peptide File Conversion (.fa to.csv)

The annotation file was in text format(tab-separated). Read the file, convert it into a *pandas* data frame, and finally save it to a CSV file named 'annotation_info.csv' as shown in Fig. 4. The problem statement lies in classifying amino acid sequences into protein families. Extract relevant features such as Sequence and ID columns from 'peptide_file.csv' and Protein families (*Defline*) and Transcript name columns from 'annotation_file.csv'. Merge these two files to make a single dataset named 'filtered_data.csv' shown in Fig. 5 that is based on common identifiers present in both files (*ID* and *Transcript name*) respectively. Protein class labels were extracted based on their occurrence count of at least 50 counts (count > = 50), the required condition to choose labels, and their corresponding sequence and id. The final dataset named 'filtered_data.csv' consists of 3824 sequence samples over 34 different protein classes as shown in Table 2.

Convert peptide data from FASTA (.fa) format to comma separated values (.csv) format by extracting gene sequences information that ensures compatibility with various data processing tools.

Table 2. Sample Class Labels and its Description

S. No.	Class Label	Significance
1	'F-box family protein'	Responsible for vegetative and reproductive growth
2	'Ankyrin repeat family protein'	Involved in mediating protein-protein interactions
3	'Disease resistance protein'	Pathogen Recognition
4	'Leucine-rich repeat family protein'	Regulates shoot and root growth
5	'Leucine-rich repeat transmembrane protein'	Innate immunity in plants
6	'Protein kinase family protein'	Respond to environmental stresses
7	'Protein kinase, putative'	Regulate biological activity
8	'Short-chain dehydrogenase/reductase family protein'	Metabolic regulation
9	'Zinc finger protein-related'	Regulate growth and Stress adaptation
10	'Zinc knuckle family protein'	Nucleic acid and zinc ion binding

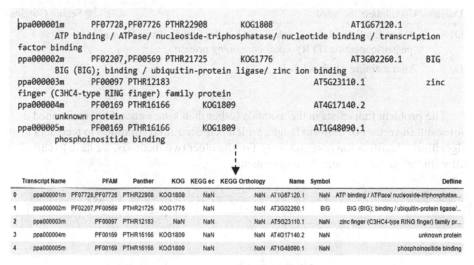

Fig. 4. Annotation File Conversion (.txt to.csv)

Fig. 5. Final Dataset (filtered_data.csv)

3.4   K-mer Encoding

When processing the Genome sequence, conversion from string format to numerical value is necessary, to form a matrix input for model training. The features of the existing sequence encoding methods are shown in Table 3.

Table 3. Sequence Encoding Methods

Encoding Method	Features
Sequential encoding	Encodes each character into a numeric value
One-hot encoding	Represents the categorical to binary value mapping in a binary vector format
K-mer encoding	Divide the sequence into K-length overlapping short sequences or segments

Table 4. Details of the Dataset Split

Dataset	Class labels included	# Search Queries
D1	'F-box family protein', 'nucleic acid binding/ribonuclease H', and 'pentatricopeptide (PPR) repeat-containing protein'	169
D2	All remaining 31 class labels	595

The problem that exists in the methods (other than *k-mer* encoding) mentioned does not result in vectors of uniform length, which is the required condition to feed data to an algorithm (classification or regression). For the other two methods, we have to curtail or fill with "*n*" or "*0*" to meet the requirement.

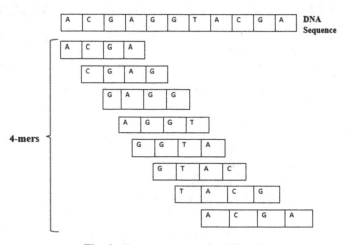

Fig. 6. *K*-mer representation [$K = 4$]

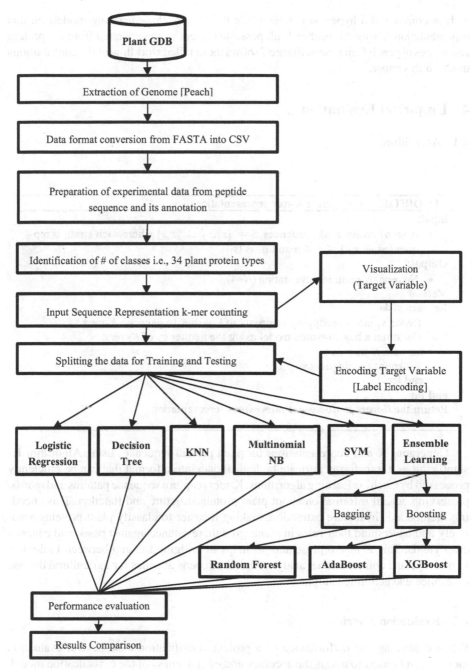

Fig. 7. Detailed Workflow

Choose the value of K, which represents the length of the subsequences (K-mers) used for encoding as shown in Fig. 6. Typically, K is set to a small value, such as 4 or

6. It is considered a hyper-parameter while training machine learning models on this representation. Using Algorithm 1, all possible 6-mers were generated from the protein sequences of peach fruit plants. Figure 7 shows the detailed workflow of the contributions made in this paper.

4 Empirical Evaluation

4.1 Algorithms

ALGORITHM 1: Generate K-mer representation

Input:
- A set of amino acid sequences, $S = \{s_1, s_2, \ldots, s_m\}$ where each string is represented as $s_i, 1 \leq i \leq n$ and $n = |s_i|$

Output:
- Numeric N-gram representation (N=4)

Method:

for each s_i do
    Divide s_i into overlapping segments of length K to generate K-mers.
    Construct a bag-of-words model using the frequency of K-mers.
    for each K-mer **do**
        count the frequency
    end for
end for
Return the K-mers in *document-term matrix* representation.

Generating K-mer representations for plant protein sequences using Algorithm 1is significant as it transforms intricate biological data into a format that can be efficiently processed by machine learning algorithms. K-mers capture sequence patterns and motifs, preserving crucial information about plant protein structure and function. This encoding method aids in feature extraction, making it easier to classify plant proteins accurately and understand their roles in plants growth, resistance against pests, and enhance crop yields. This K-mer representations helps in bridging the gap between biological knowledge and computational analysis, and thus opens avenues for agricultural disease resistance and genomic analytics.

4.2 Evaluation Metrics

When evaluating the performance of a protein classification task, several evaluation metrics can be used to assess the accuracy and effectiveness of the classification model. The proposed method has been evaluated using the following metrics.

Confusion Matrix: It can be used to derive other evaluation metrics such as accuracy, precision, and recall.

Accuracy: Accuracy measures the proportion of correctly classified protein instances over the total number of instances in the dataset. It provides an overall measure of the model's correctness.

$$Accuracy = TP + TN/(TP + TN + FP + FN) \tag{1}$$

Precision: Precision represents the proportion of true positive predictions (correctly classified positives) over the total positive predictions. It indicates the model's ability to avoid false positives.

$$Precision = TP/(TP + FP) \tag{2}$$

Recall (*a.k.a.* Sensitivity or True Positive Rate): Recall measures the proportion of true positive predictions over the total actual positive instances in the dataset. It reflects the model's ability to correctly identify positive instances.

$$Recall = TP/(TP + FN) \tag{3}$$

F1-score: It is the harmonic mean of *precision* and *recall*. It provides a balanced measure, which is useful when the dataset has imbalanced classes.

$$F1 - score = 2 * (precision * recall)/(precision + recall) \tag{4}$$

4.3 Experimental Evaluation Using Entire Dataset (with 34 classes)

The dataset shown in Fig. 5 is used to train the models. The detailed dataset preparation is mentioned in Sect. 3.3. The dataset consists of 764 sequence samples over 34 different protein classes as sample classes. Different machine learning algorithms were considered for building the classification model.

It shows that the Multinomial Naïve Bayes model outperformed other models while observing the results shown in Fig. 8. Analyzing various classifiers, it is observed that three class labels namely '*F-box family protein*', '*nucleic acid binding/ribonuclease H*', *and* '*pentatricopeptide (PPR) repeat-containing protein*' were assigned to most of the instances. Therefore, the dataset is split into two parts and has been trained separately such that one dataset comprises 3 classes whereas the other dataset comprises of remaining 31 classes. Due to the rise in classification errors of those 3 classes, it was decided to divide the dataset into two parts called **D1** and **D2** (Table 4).

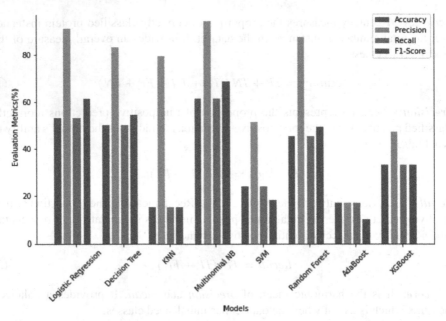

Fig. 8. Evaluation scores of 7 ML models considered for experimentation

4.4 Performance Evaluation on D1 and D2

It is noted that the Support Vector Machine (SVM) outperformed on dataset **D1** that addresses 3-class classification task, while Multinomial NB outperformed on dataset **D2** that addresses 31-class complex classification task from Tables 5 and 6 respectively. Assessing complex protein classes entails rigorous analysis of their structure, function, and interactions. This process involves data integration, and computational modeling that enables deeper insights into plant biological systems.

Table 5. Results obtained on dataset D1

Models	Accuracy	Precision	Recall	F1-score
Logistic Regression	63	81	63	60
Decision Tree	66	80	66	62
KNN	12	26	12	11
Multinomial NB	21	92	21	22
SVM*	**76**	**70**	**76**	**73**
Random Forest	56	77	56	46
AdaBoost	53	35	53	38
XGBoost	64	79	64	59

Table 6. Results obtained on dataset D2

Model	Accuracy	Precision	Recall	F1-score
Logistic Regression	50	90	50	55
Decision Tree	50	88	50	55
KNN	15	81	15	21
Multinomial NB*	66	90	66	72
SVM	27	57	27	25
Random Forest	42	83	42	47
AdaBoost	16	21	16	12
XGBoost	34	51	34	34

4.5 Tools and Libraries Used

A range of Python libraries and toolkits were harnessed for the implementation of the presented *peach* plant classification models. Particularly, the *sci-kit-learn* library was employed for modeling algorithms including Logistic Regression, Decision Tree, KNN, multinomial NB, SVM, Random Forest, AdaBoost, and XGBoost on the prepared dataset. *BioPython* library was employed in handling biological data and gene sequences from the peach genome. It provided functionalities for gene sequence manipulation, feature extraction, and data preprocessing, ensuring the effective integration of biological data and knowledge into the machine-learning pipeline.

5 Conclusions

Plant protein classification using *K*-mer encoding involves representing protein sequences as fixed-length feature vectors based on the frequencies of subsequences. This encoding method provides a compact and informative representation of protein sequences, enabling the application of ML algorithms for classification. Additionally, experimenting with different *K*-mer lengths would improve the classification accuracy of the models. It is observed that the *multinomial NB* outperforms with an F-score of 72% when evaluated on the dataset containing 31 features while *SVM* reports 73% on the dataset containing 3 features. Achieving a remarkable F-score of 72% in a complex 31-class classification problem demonstrates robust model performance, effective feature engineering, and rigorous model tuning. These developments contribute to the improvement of classification accuracy and the understanding of plant protein functions and interactions, which have significant implications in the precision agriculture domain. Further attempts could be made to interpret the model's prediction output through explainable artificial intelligence (EAI) capabilities. Explaining the output of the ML models is still an open area for research as it is considered challenging to interpret the output of an extremely complex model.

References

1. Öncül, A.B., Çelik, Y.: A hybrid deep learning model for classification of plant transcription factor proteins. SIViP **17**, 2055–2061 (2023)
2. Nedyalkova, M., Vasighi, M., Azmoon, A., Naneva, L., Simeonov, V.: Sequence-based prediction of plant allergenic proteins: machine learning classification approach. ACS Omega **8**(4), 3698–3704 (2023). https://doi.org/10.1021/acsomega.2c02842
3. Upadhyaya, S.R., et al.: Evaluating Plant Gene Models Using Machine Learning. Plants **11**(12), 1619 (2022). https://doi.org/10.3390/plants11121619
4. Yadav, A.K., Singla, D.: VacPred: Sequence-based prediction of plant vacuole proteins using machine-learning techniques. J. Biosci. **45**, 1–9 (2020)
5. Simon, O.A., et al.: K-mer-based machine learning method to classify LTR-retrotransposons in plant genomes. PeerJ **9**, e11456 (2021)
6. Warin, W., et al. Ensemble of multiple classifiers for multilabel classification of plant protein subcellular localization. Life **11.4**, 293 (2021)
7. Guo, Y., Hou, L., Zhu, W., Wang, P.: Prediction of Hormone-Binding Proteins Based on K-mer Feature Representation and Naive Bayes. Front. Genet. **12**, 797641 (2021)
8. Juneja, S., Dhankhar, A., Juneja, A., Bali, S.: An approach to DNA sequence classification through machine learning: DNA sequencing, K-Mer counting, thresholding, sequence analysis. Int. J. Reliable Qual. E-Healthcare (IJRQEH) **11**(2), 1–15 (2022)
9. Sangphukieo, A., Laomettachit, T., Ruengjitchatchawalya, M.: Photosynthetic protein classification using genome neighborhood-based machine learning feature. Sci. Rep. **10**(1), 7108 (2020)
10. Gotoh, O., Morita, M., Nelson, D.R.: Assessment and refinement of eukaryotic gene structure prediction with gene-structure-aware multiple protein sequence alignment. BMC Bioinform. **15**(1), 1–13 (2014)
11. http://plantgdb.org/PeGDB/ - Prunus persica [Peach] Genome. Accessed 14 June 2023
12. Pan, J., et al.: DWPPI: a deep learning approach for predicting protein-protein interactions in plants based on multi-source information with a large-scale biological network. Front. Bioeng. Biotechnol. **10**, 807522 (2022)
13. Li, L.-P., Zhang, B., Cheng, L.: CPIELA: computational prediction of plant protein-protein interactions by ensemble learning approach from protein sequences and evolutionary information. Front. Genet. **13**, 857839 (2022)

Deep CNN Based Alzheimer Analysis in MRI Using Clinical Dementia Rating

Abhishek Saigiridhari[✉], Abhishek Mishra, Aarya Tupe, Dhanalekshmi Yedurkar, and Manisha Galphade

MIT School of Engineering, MIT-ADT University, Pune, Maharashtra, India
abhisheksaigiridhari@gmail.com

Abstract. Globally, Neurological disorders are a major health concern affecting a population of billions worldwide. There's a need for accurate and timely diagnosis of brain disorders to improve patient outcomes and revolutionize the field of medicine with the help of technology. For this, the integration of deep learning models with MRI (structural and functional) images presents a promising approach for the detection of brain disorders like Alzheimer's disease. Our Research aims to develop and evaluate deep learning models for detecting Alzheimer's disease using the Oasis dataset, a popularly used data set of neuroimaging and processed imaging data, for brain images of Alzheimer patients. There were 2 types of images i.e. the Raw and FSL-SEG (preprocessed) gifs. The models were developed using multiple Convolution layers and a Non-linear activation function (Sigmoid) for binary classification. Early stopping on loss helped prevent overfitting, and a batch size of 75 was used for faster convergence. We generated an accuracy of 90% on the FSL-SEG MRI images whereas the RAW images resulted in an accuracy of 83%. With a value of 0.79 in Area Under the Curve, The CDR (Clinical Dementia Rating) as well as MMSE (Mini Mental State Examination) were main factors which interlinked the images with occurence of Alzheimer.

Keywords: Alzheimer · Clinical Dementia Rating (CDR) · MRI · OASIS Dataset · FSL-SEG · Raw Images · CNN

1 Introduction

It is a well-known fact that neurological disorders are the number one cause of disability in the world. More than a billion individuals globally suffer from some form of neurological disorder. Alzheimer affects a significant number of individuals worldwide, yet a large proportion of cases remain undiagnosed. Currently, there is no definitive medical test available for the precise identification of this disorder. Consequently, individuals with alzheimer are typically diagnosed only when their symptoms become noticeable and impact their daily lives. These symptoms encompass memory loss, difficulty concentrating, communication and language confusion, visual perception impairment, and a decline in reasoning and judgment abilities. The question arises: Can neural networks contribute to the early detection of dementia and Alzheimer's disease? If so, how can

R. Muthalagu et al. (Eds.): CINS 2023, CCIS 1978, pp. 105–116, 2024.
https://doi.org/10.1007/978-3-031-48984-6_9

early diagnoses assist individuals in postponing the life-altering symptoms they may eventually face? Furthermore, how might the concepts of CNNs or transfer learning models be applied to aid in the treatment of other neurological disorders? The integration of deep learning models with MRI images presents a promising approach to achieving this goal, as it allows for the identification of specific patterns of brain activity indicative of various conditions. By leveraging this technology, healthcare professionals can provide timely and personalized treatment options, ultimately leading to improved quality of life for those affected by neurological disorders. Hence we have made use of the OASIS dataset to detect Alzheimer's by analyzing the RAW as well as the FSL-SEG images. We have also done an exploratory data analysis on the cross-sectional.csv data which contains relevant information about the image data.

2 Related Works

In recent years, there has been a growing interest in using deep neural networks for the prediction and classification of brain disorders, including Alzheimer's disease and epilepsy. Several researchers have explored the application of computational modeling and machine learning techniques to analyze various neuroimaging modalities and predict susceptibility to these disorders.

Garnet et al. [1] focused on using deep neural networks to predict Alzheimer's disease using the OASIS dataset. The study utilized convolutional neural net works (CNNs) and the CapNet approach to classify individuals into different categories based on features such as MR Delay, Gender, Age, Education, and cognitive scores. The results demonstrated the potential of deep neural networks in accurately predicting Alzheimer's disease.

In Shoeibi et al. [2], multimodal machine learning techniques were employed for the classification of post-traumatic seizures using dMRI, EEG, and fMRI data. The study introduced the IDSF algorithm, which captured shared and unique information from multiple modalities. The findings suggested that fMRI alterations in the inferior temporal gyrus could serve as potential biomarkers for post-traumatic epilepsy.

A study Basheer et al. [6] focused on utilizing resting-state fMRI connectivity and machine learning methods to predict seizure susceptibility in individuals with epilepsy. The study employed graph theory based connectivity measures and support vector machine (SVM) classifiers to achieve high accuracy in predicting seizure susceptibility. The findings highlighted the importance of connectivity measures in the brain cortex for accurate prediction.

Another research paper Basaia et al. [4] developed and validated a deep learning algorithm for the automated classification of Alzheimer's disease and mild cognitive impairment (MCI) using MRI scans. The study employed convolutional neural networks (CNNs) on 3D T1-weighted images and achieved high accuracy in distinguishing between different diagnostic groups. In a similar vein, Aaraji et al. [5] explored the use of deep CNNs for the automatic classification of Alzheimer's disease based on brain MRI data. The authors discussed the challenges of AD detection and proposed a CNN-based approach that extracted discriminative features directly from raw data. The experimental results demon strated promising accuracy rates for AD classification.

Lastly, Kruthikaa et al. [3] proposed a computer-aided diagnosis system for the early detection of Alzheimer's disease using content-based image retrieval (CBIR) techniques. The study applied 3D Capsule Networks, 3D CNNs, and pre-trained 3D autoencoders to facilitate relevant image retrieval and training for AD detection.

3 Methodology

3.1 Characteristics of the MRI Images

The OASIS cross-sectional dataset is composed of MRI scans from a total of 416 individuals, whose ages range from 18 to 96 years old. Each participant underwent between 3 and 4 T1-weighted MRI scans in a single scan session. Within this dataset, there is a subgroup of 100 individuals over the age of 60 who have received a diagnosis of very mild to moderate Alzheimer's disease. Along with the main dataset, an additional dataset of 20 individuals who have not been diagnosed with dementia and who were imaged during a second visit within 90 days of their first session is also provided. The OASIS dataset contains two types of image files: Raw MRI images and FSL-SEG images.

Raw MRI images (Fig. 1b) are the original, unprocessed images captured by the MRI scanner. They contain information about the structure and composition of the brain but also include noise and artifacts.

FSL-SEG images (Fig. 1a), on the other hand, are processed and segmented MRI images. They have been run through a software tool called FSL (FMRIB Soft ware Library) which segments the images into different tissue types such as gray matter, white matter, and cerebrospinal fluid (CSF). These images provide a more detailed representation of the brain structure and are commonly used for brain image analysis.

Oasis Cross-Sectional Data: The Oasis Dataset contains the following Important attributes for each subject:

1. M/F: The gender of the subject (M for male, F for female).
2. Age: The age of the subject at the time of the MRI scan.
3. MMSE: The Mini-Mental State Examination score, a measure of cognitive function.
4. CDR: The Clinical Dementia Rating score, a measure of the severity of dementia.
5. eTIV: The estimated total intracranial volume, a measure of brain size.
6. nWBV: The normalized whole-brain volume, a measure of brain atrophy.
7. ASF: The atlas scaling factor, a measure of the size of the brain.
8. Delay: The delay between the baseline and follow-up MRI scans, in months.

These attributes are used in the Oasis Dataset to study the effects of aging and dementia on brain structure and function.

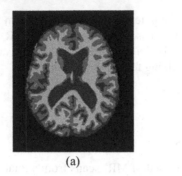
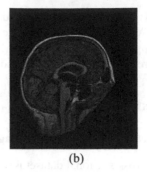

(a)   (b)

Fig. 1. a. FSL-SEG Image of Brain, b. RAW MRI Image of Brain

3.2 Analysis of the OASIS Cross-Sectional Dataset

From Descriptive statistics, we see that:

- The average age of the participants is 51.4 years, and the age range is from 18 to 96 years old.
- The MMSE variable ranges from 14 to 30, with an average of 27.1, indicating that the participants, on average, have good cognitive function.
- The clinical dementia rating (CDR) variable ranges from 0 to 2, with an average of 0.3, indicating that most participants do not have significant dementia.
- The estimated total intracranial volume (eTIV) ranges from 1123 to 1992 cubic centimeters, with an average of 1481.9.
- The normalized whole brain volume (nWBV) ranges from 0.644 to 0.893, with an average of 0.7917.
- The atlas scaling factor (ASF) ranges from 0.881 to 1.563, with an average of 1.1989.
- The Delay variable only has 20 non-missing values, with a minimum of 1 and a maximum of 89. The average delay is 20.55.

3.3 Understanding MRI Image Data: Anatomy and Interpretation

We have specifically considered the FSL—SEG images to understand the features and interpret the anatomy of the brain during Dementia. Here the FSL SEG Image denoted the Coronal plane of the Brain.

A coronal plane is an anatomical plane used to divide the human body or brain into distinct sections. It is a vertical plane that separates the body or brain into anterior (front) and posterior (back) portions. In the context of the brain, a coronal section refers to a slice made perpendicular to the sagittal plane (which divides the brain into left and right halves) and parallel to the face.

There were significant differences in the features of a patient with CDR >1 and with CDR – 0. CDR as mentioned is the Clinical Dementia Rating, which defines whether the patient's brain is : Demented: 2, Moderately Demented: 1, Mildly Demented-0.5, Normal -0.

The features that were different are listed below:

1. Lateral Ventricles of the Brain:

   In Alzheimer's disease, the volume of the lateral ventricles tends to increase. This enlargement is believed to be due to the loss of brain tissue, including the shrinking of structures surrounding the ventricles (Fig. 2).

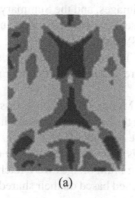 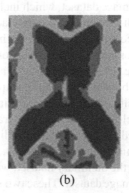

(a)   (b)

Fig. 2. a) Normal size of Lateral Ventricle. b) Enlarged Lateral Ventricle

2. Size of the Brain:

   Alzheimer's disease is characterized by progressive brain atrophy, which means a reduction in the size of the brain. As the disease progresses, the brain tends to shrink, and this shrinkage is most pronounced in the hip pocampus.

3. White Matter region in the Brain:

   Several studies have shown that individuals with Alzheimer's disease exhibit reduced white matter volume and decreased integrity of white matter tracts compared to healthy individuals. These changes can be observed using imag ing techniques such as magnetic resonance imaging (MRI). We see that in the above Fig. 3, there is a significant decrease in the white matter region in the brain. This surely indicates that the CDR is 2.

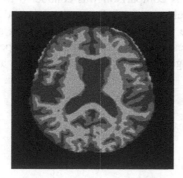

Fig. 3. Shrinked size of the Brain, due to loss of Brain Cells

These features have been used to determine whether or not the Patient's brain is demented or normal.

3.4 Architecture and Diagram

1. Integration of Image and Summary Data: The Oasis-1 dataset consists of two formats: The image dataset, which includes FSL and RAW images, and the Summary dataset, which contains CSV data. Each patient in the dataset is associated with a subject ID that corresponds to both the image data and the CSV data.
2. Exploratory Data Analysis on Summary Dataset: To gain a deeper under standing of the data. Various graphs and visualizations were generated to uncover patterns, distributions, and relationships within the data.
3. Training and Testing a Simple Classification Model: A simple classification model(RandomForest) was trained and tested using the Summary dataset. This model served as a baseline to establish initial performance benchmarks.
4. Fetching Labels and Images: The labels (Diagnosis) for each patient were obtained from the Summary dataset, while the corresponding brain images were fetched from the Image dataset. These two sets of data were matched based on their shared Subject ID, enabling the creation of a cohesive dataset.
5. Creating a data frame: The Subject ID, brain image, and Diagnosis information were combined and stored as a data frame. This consolidated data frame served as the input for further analysis and model development.
6. Utilizing Deep Learning Models: Deep learning models (Convolutional Neural Networks (CNN) and Residual Networks (ResNet)), were constructed using the created data frame as the input.
7. Model Building and Evaluation: The CNN and ResNET models were trained and evaluated. The training involved optimizing model parameters using the labeled data, while evaluation focused on assessing the models' performance on unseen data. This process ensured the models' ability to generalize and make accurate predictions.
8. Obtaining Results: Various metrics, such as accuracy and loss, were obtained to quantify the performance of the trained models. Additionally, the Receiver Operating Characteristic (ROC) curve was plotted to assess the models' ability to distinguish between different classes.
9. Comparative Analysis: A comparison was conducted to evaluate the performance of a classification model using both Summary data and Image data. The analysis demonstrated that there were only slight differences in accuracy between the two types of data. This suggests that the classification model performed effectively across both Summary and Image data, indicating its robustness and reliability (Fig. 4).

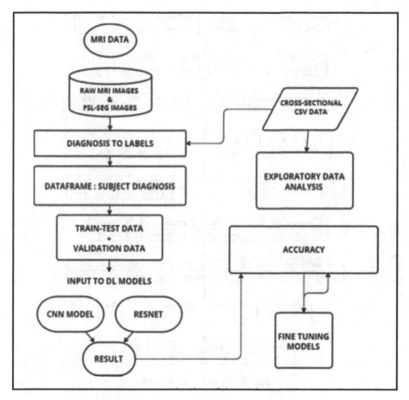

Fig. 4. Represents the Flow of the pipeline

3.5   Model Structure: Network Architecture and Layers

Figure 5 represents the Model Architecture, The architecture of the CNN Model consists of 3 Convolutional layers with decreasing numbers of filters (100, 50, and 25), followed by max-pooling layers.

The convolutional layers use the sigmoid activation function and are set up to preserve the spatial dimensions of the input by using the same padding, while also applying a stride of (10,10) in the first layer and (5,5) in the second layer. The max-pooling layers use a 2×2 pooling window with valid padding in the first two layers and a 1×1 pooling window with valid padding in the third layer.

After the max-pooling layers, the feature maps are flattened into a 1D vector and fed into a dense layer with a single neuron, which uses the sigmoid activation function to output a binary classification prediction.

The model is compiled using binary cross-entropy loss and the Adam optimizer. During training, early stopping is applied to prevent overfitting.

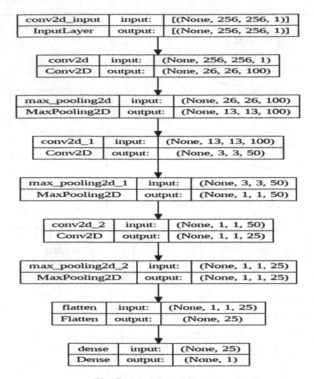

Fig. 5. Model Architecture

4 Result and Analysis

4.1 Model Training and Accuracy

After training the CNN model on both FSL-SEG and RAW images (Training set: 64%, Testing set: 20%, Validation set: 16%), the resulting accuracies were 90.62% and 83.12%, respectively. These accuracies indicate the percentage of correctly classified samples by the model.

We evaluated the model's performance using metrics such as accuracy, AUC, loss and Various Activation Functions (Fig. 6).

SR.NO	ACTIVATION FUNCTIONS	ACCURACY	LOSS	AUC VALUE
1	SIGMOID	90.62	0.3820	0.79
2	TANH	74.41	1.9765	0.61
3	RELU	72.73	2.2111	0.53
4	LEAKY-RELU	74.51%	2.1239	0.47
5	ELU	54.55%	2.466	0.59
6	SWISH	72.73%	4.4273	0.41

Fig. 6. Comparison of Different Activation Functions

Sigmoid Activation: The sigmoid activation function, known for producing probabilities, achieved an accuracy of 90.62%. This indicates that the model performed exceptionally well in binary classification tasks. However, the AUC value of 0.79 suggests that the model's ability to distinguish between classes is pretty good. The moderately high AUC value indicates that the sigmoid activa tion function was the optimal choice for this specific problem. The achieved loss of 0.382 suggests a relatively good fit to the training data.

ReLU Activation: ReLU, a popular activation function due to its simplicity and effectiveness, yielded an accuracy of 72.73%. The AUC value of 0.53 indicates a moderate ability to discriminate between classes.

Tanh Activation: The hyperbolic tangent (tanh) activation function resulted in an accuracy of 72.73%, similar to ReLU. The AUC value of 0.61 indicates a better ability to differentiate between classes compared to ReLU.

LeakyReLU Activation: With the LeakyReLU activation function, the model achieved an accuracy of 74.51%, slightly higher than ReLU.

ELU: ELU, another variant of the ReLU function, resulted in an accuracy of 54.55%, which is the lowest among the activation functions tested.

Swish: Swish, a recently proposed activation function, achieved an accuracy of 72.73%, similar to ReLU and tanh.

In summary, the sigmoid activation function yielded the highest accuracy of 90.62% with a good AUC.

4.2 Area Under the Curve Analysis

An AUC of 0.79 suggests that the model's ability to classify positive and neg ative instances is moderately good. However, it's important to note that the interpretation of an AUC value depends on the context and the specific problem being addressed. In some cases, an AUC of 0.79 might be considered quite good, while in other cases, it may be less desirable depending on the requirements and expectations of the task at hand (Fig. 7).

There are 27 true negatives, meaning the model correctly predicted 27 sam ples as negative. There are 6 true positives, indicating that the model correctly predicted 6 samples as positive. However, there are 14 false positives, where the model predicted 14 samples as positive when they were actually negative. Additionally, there are 8 false negatives, indicating that the model predicted 8 samples as negative when they were actually positive (Fig. 8).

We can clearly see that In Fig. 9a and b it has correctly predicted as a Demented Brain and Non-Demented Brain. There may be many factors that the neural network would've captured like the lateral ventricles, significant shrink in the size of brain tissues, and reduction in white matter in the brain.

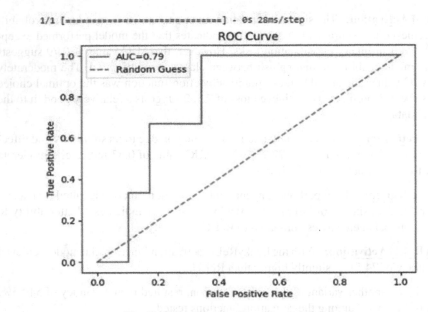

Fig. 7. Area under the Curve for the Model

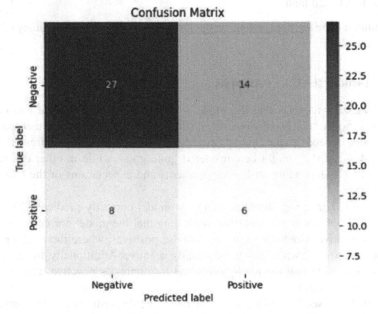

Fig. 8. Confusion Matrix

OAS1_0039 Classification: 1   OAS1_0037 Classification: 0

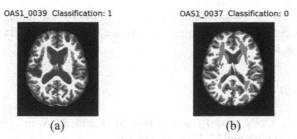

(a)   (b)

Fig. 9. a) Demented Brain. b) Non-Demented Brain.

4.3 Limitations and Discussions

In this research study, a proposed CNN model shows an exceptionally good result on both FSL-SEG and RAW images, the resulting accuracies were 90.62% and 83.12%, respectively. The model performed well at pattern recognition, as evidenced by its high precision in classifying FSL-SEG images and accuracy of 83.12% for RAW images. The slight difference in accuracy between the two datasets could be explained by variations in image quality, preprocessing, and content.

This study has the following limitations:

1. The availability of data was limited, with only 411 out of 436 images from the FSL-SEG dataset and 383 images from the RAW dataset utilised.
2. Due to the absence of labelled data and the requirement for accurate labelling, Data Augmentation was not pursued. The limited availability of labelled data prevented us from utilising this approach to potentially enhance model performance and generalizability.
3. Our study did not account for any clinical trials or interventions that patients may have undergone prior to the MRI images being taken. These interventions, such as medication or therapies,could impact the imaging results and introduce confounding factors that were not addressed in our analysis.
4. Our analysis focused on the detection of specific features, such as the hippocampus, corpus callosum trunks, and ventricles. While this approach may facilitate easier analysis and detection of abnormalities in these regions, it overlooks potential information contained in other areas of the brain. Consequently, our findings may not represent a comprehensive assessment of the entire brain structure.

The future scope of this research is quite reassuring and encouraging, and there is still much room for improvement in the field of brain disorder detection using deep learning techniques and MRI images.

As far as the medical field advancement is concerned, this research could also help in exploring and understanding other diseases like Parkinson's disease, epilepsy, and brain tumours by using similar methods, algorithms, and help in building a comprehensive disease diagnostic tool for healthcare practitioners. Adding more diverse populations and a wider variety of brain disorders to the dataset could also make the system more generalizable and increase its usefulness in the real world.

Future research should strive to address these limitations to obtain a more comprehensive understanding of the relationships between MRI images and clinical factors in neuroimaging studies.

5 Conclusion

After training the CNN model on both FSL-SEG and RAW images, the result ing accuracies were 90.62% and 83.12%, respectively.

In summary, the sigmoid activation function yielded the highest accuracy of 90.62% with a good AUC value.

The current state of Alzheimer's diagnosis presents a significant challenge in the medical field. It is estimated that more than half of individuals living with dementia remain undiagnosed, highlighting the need for improved detection methods. Presently, there is no single medical test that can definitively identify dementia with certainty.

Deep learning models could also be used to detect brain disorders, which could lead to better treatment options, lower medical expenses, and higher patient satisfaction. By incorporating these models into national healthcare systems and encouraging their widespread adoption among healthcare professionals, govern ments can take advantage of these advantages.

Future research should strive to address these limitations to obtain a more comprehensive understanding of the relationships between MRI images and clinical factors in neuroimaging studies.

References

1. Garner, R., Ghariq, E., Vasavada, M.M., Singh, K., Carney, P.R.: Machine learning model to predict seizure susceptibility from resting-state fMRI connectivity. Front. Neurol. **12**, 715929 (2021)
2. Shoeibi, A., Khosravi, A., Wang, X.: An overview of deep learning tech niques for epileptic seizures detection and prediction based on neuroimaging modalities: methods, challenges, and future works. Front. Neurosci. **15**, 650669 (2021)
3. Kruthikaa, K.R., Rajeswari, B., Maheshappa, H.D.: CBIR system using Capsule Networks and 3D CNN for Alzheimer's disease diagnosis. Alzheimer's Dis ease Neuroimaging Initiative1 (2021)
4. Basaia, S., et al.: Automated classification of Alzheimer's disease and mild cognitive im pairment using a single MRI and deep neural networks. Alzheimer's Disease Neu roimaging Initiative1 (2021)
5. Aaraji, Z., Abbas, H.H.: Automatic Classification of Alzheimer's Disease using brain MRI data and deep Convolutional Neural Networks. IEEE Access **9**, 45241–45252 (2021)
6. Basheer, S., Bhatia, S., Sakri, S.B.: Computational Modeling of Demen tia Prediction Using Deep Neural Network: Analysis on OASIS Dataset (2021)
7. Christensen, D.V., et al.: 2022 roadmap on neuromorphic computing and engineering. Neuromorphic Comput. Eng. **2**(2), 022501 (2022)
8. https://fsl.fmrib.ox.ac.uk/fsl/fslwiki/
9. https://keras.io/
10. https://www.oasis-brains.org/data

Disaster Tweets Classification for Multilingual Tweets Using Machine Learning Techniques

Tanya Koranga$^{(\boxtimes)}$, Raju Hazari$^{(\boxtimes)}$, and Pranesh Das

Department of Computer Science and Engineering National Institute of Technology Calicut, Kerala, India
tanyakoranga1208@gmail.com, rajuhazari@nitc.ac.in

Abstract. Natural disasters have dire consequences for communities, leading to loss of life, property destruction and environmental devastation. Effective disaster response necessitates prompt and coordinated actions. Social media platforms, particularly Twitter, have emerged as invaluable assets in disaster management. With an enormous daily influx of tweets, Twitter data presents an opportunity to gain valuable insights for tracking and responding to disasters. However, sifting through the vast volume of regular content to identify relevant tweets poses a significant challenge. Furthermore, the global nature of Twitter introduces an added layer of complexity with tweets in different languages. Recent advancements in deep learning techniques provide promising solutions for addressing this challenge, enabling the identification of disaster-related information from multilingual tweets. This research proposes a comprehensive approach that leverages machine learning and deep learning models to accurately classify disaster-related tweets in multiple languages, including English, Hindi, and Bengali. The study evaluates the performance of seven Machine Learning classifiers, including Naive Bayes, Logistic Regression, Random Forest, Support Vector Machine, K-Nearest Neighbors, Gradient Boosting, Decision Tree and few Deep Learning models such as LSTM, BiLSTM, BiLSTM with CNN, BERT and DistilBERT. After conducting a thorough evaluation of multiple models, it is evident that BERT and DistilBERT stand out as the top performers, consistently exhibiting exceptional accuracy and delivering consistent results across diverse language contexts.

Keywords: Tweets classification · LSTM · BiLSTM · Transformers · BERT

1 Introduction

Disasters have the potential to cause widespread devastation, impacting communities at various scales, from individual households to entire nations. Whether they are natural disasters like floods, hurricanes, and earthquakes, or human-caused disasters such as industrial accidents and terrorist attacks, their effects on people's lives and the environment can be profound. During and after a disaster, effective communication and information sharing are crucial for allocating

© The Author(s), under exclusive license to Springer Nature Switzerland AG 2024
R. Muthalagu et al. (Eds.): CINS 2023, CCIS 1978, pp. 117–129, 2024.
https://doi.org/10.1007/978-3-031-48984-6_10

resources and coordinating response efforts. Social media platforms like Twitter have emerged as invaluable tools for sharing real-time information about disasters, allowing individuals to alert authorities about emerging situations and receive updates on rescue operations.

However, the vast amount of data generated on social media during disasters can be overwhelming, presenting a significant challenge in identifying and prioritizing the most critical needs. Compounding this challenge is the presence of tweets in different languages, adding another layer of complexity. The multilingual nature of Twitter further complicates the process of extracting relevant information from a diverse range of languages.

Machine learning algorithms, such as logistic regression, decision trees, and Support Vector Machines (SVMs), have been extensively employed for tweet classification. These algorithms leverage various statistical techniques to learn patterns and relationships in the data. They can handle high-dimensional datasets and are effective in cases where the relationships between features and classes are relatively simple.

On the other hand, deep learning algorithms, have gained significant attention in recent years due to their ability to automatically learn hierarchical representations of data. Deep learning models, such as Bidirectional Long Short-Term Memory (BiLSTM) networks, Convolutional Neural Networks (CNNs), and Transformer models, have shown remarkable performance in tasks involving sequential and textual data. Deep learning models, including BiLSTMs and BERT (Bidirectional Encoder Representations from Transformers), are highly effective for tweet classification. BiLSTMs excel at capturing long-term dependencies and sequential patterns, while BERT, a transformer-based model, has revolutionized natural language understanding tasks.

To address multilingual challenges, machine learning and deep learning techniques can be extended to handle different languages. By training models on diverse language datasets and leveraging techniques like multilingual embeddings or language independent preprocessing, it is possible to classify tweets in languages like Hindi or others. By combining these techniques, disaster-related tweets can be efficiently classified, aiding in resource allocation and coordination efforts during and after disasters, regardless of the language in which they are written. Among the various algorithms employed in our study, it was evident that BERT and DistilBERT emerged as the top performers.

2   Related Works

Various strategies and approaches have been proposed to develop models for classifying social media data. These models aim to accurately categorize and represent the diverse content found in social media platforms. In [3], authors used four classifiers: SVM, KNN, Naïve Bayes, and XGBoost. SVM and XGBoost performed well with accuracies of 80% and 78% respectively, while KNN suffered overfitting (99% accuracy) and Naïve Bayes performed poorly (65%). In [21], the model is trained using two algorithms named as Multinomial Naive Bayes and AdaBoost Algorithm where Multinomial Naive Bayes gives an accuracy of 87.15% and AdaBoost has about 83% of accuracy.

Deep learning models were also explored in disaster tweet classification. In a specific study [16], CNN, LSTM, BiLSTM and BiLSTM with attention mechanism were investigated. The findings demonstrated that CNN performed exceptionally well with Hindi tweets, while BLSTM with attention mechanism, utilizing crisis word embeddings, yielded superior results for English datasets. In [7], authors compared different embeddings and preprocessing techniques for deep learning-based classification of disaster tweets. CNN with pre-trained BERT achieved the best results on a dataset without stop words removal and stemming, while Skip-gram performed consistently well on both datasets. The work showed that stop words removal may decrease model performance due to reduced dataset size and potential changes in text meaning. It highlighted the importance of larger datasets for improved performance.

In [13], authors compared conventional machine learning classifiers (SVM, RF, LR, KNN, Naive Bayes, Gradient Boost, Decision Tree) with deep learning classifiers (CNN, LSTM, GRU, Bi-GRU, GRU-CNN) for disaster-related tweet classification. The Gradient Boost (GB) classifier outperformed others, achieving high F1-scores (0.79, 0.80, 0.70, and 0.67) for Hurricane, Earthquake, Flood, and Wildfire events respectively. Deep neural network models showed superior performance (F1-scores: 0.61 to 0.88) compared to conventional ML classifiers (F1-scores: 0.16 to 0.80).

The problem of identifying multi-modal informative tweets during crises is handled in [17], where text and images play a crucial role for humanitarian organizations. A novel approach leveraging fine-tuned BERT and DenseNet is proposed to capture both textual and image content in tweets. The method is evaluated on diverse disaster datasets including Hurricane Harvey, Hurricane Irma, Hurricane Maria, California Wildfire, and Earthquake. Various performance metrics such as ROC Curve, accuracy, precision, recall, and F1-score are employed. Results demonstrate that the proposed method outperforms state-of-the-art approaches. In 2021, a sentiment aware contextual pipeline model named SentiBERT-BiLSTM-CNN was proposed [25]. SentiBERT to obtain sentiment aware contextual embeddings, a BiLSTM layer for sequential modeling and a CNN for feature extraction. SentiBert is a variant of BERT that effectively captures compositional sentiment semantics. The proposed model demonstrates superior F1 score than other existing models (CNN, BiLSTM, SentiBERT, BiLSTM-CNN, word2vec-BiLSTM-CNN, fastText-BiLSTM-CNN etc.).

Many researchers and organizations have utilized BERT for various text classification tasks, including disaster-related tweet classification. The use of BERT in [18] aims to improve the identification of earthquake-related tweets. BERT model achieved an impressive accuracy of 82.55%, further reinforcing its effectiveness in earthquake tweet classification. The findings highlight the potential of BERT and similar deep learning models for handling text classification tasks in dynamic and evolving disaster environments. Similarly, in [15], authors fine-tuned the BERT model with carefully selected hyperparameters, resulting in an impressive overall F1 score of 0.8867. This demonstrates the effectiveness of this approach in accurately classifying the tweets, leveraging the powerful language representation capabilities of BERT.

In [9], authors introduces a genetic algorithm (GA-CNN) to optimize 20 hyperparameters of a convolutional neural network. Results from seven experiments demonstrate the importance of a larger number of hyperparameters and layer-specific values. The GA-CNN outperforms other methods in terms of F1-scores during training and testing. In other research [5] in-depth analysis of disaster prediction is done for Twitter data using various word embeddings. The experimental findings indicate that contextual embeddings yield the highest accuracy for this task. Additionally, the study demonstrates the superiority of deep neural network models over traditional machine learning methods. In the study conducted by researchers in [24], the classifiers CNN, LSTM, and RNN were employed. The optimization of these models' performance involved utilizing the validation data to fine-tune their respective hyperparameters. By iteratively training the models with 10 new samples, the F1 score was evaluated on the validation set, mimicking the gradual labeling process. After identifying the optimal hyperparameters for each classifier, their performance was assessed using the testing set. The use of an interactive learning framework allowed for real-time training of text relevance classifiers, empowering users to improve situational awareness by accurately classifying incoming data. There are more notable work related to Disaster Tweets Classification is reported in [4, 6, 8, 10–12, 14, 19, 22, 23, 26].

3   Proposed Methodology

In this work, we utilized four datasets for our study: two English datasets, one manually created Hindi dataset, and one manually created Bengali dataset. These manually created datasets are valuable resources specifically designed for disaster tweet classification in Hindi and Bengali languages. After collecting the datasets, we then proceeded with data preprocessing, where we performed cleaning and preparation tasks on each dataset individually. Following that, we employed feature extraction techniques to obtain non-sequential vectors suitable for machine learning classifiers. These vectors capture important characteristics of the text data and enable effective representation in the classification process. Additionally, for deep learning classifiers, we utilized sequential feature extraction methods to capture the sequential nature of the data. By incorporating both non-sequential and sequential feature extraction approaches, we aimed to leverage the strengths of each in enhancing the performance of our classification models. The methodology employed for classifying disaster tweets is described below in Fig. 1.

3.1   Dataset Description

This study encompasses the analysis of four datasets, comprising two English datasets, one Hindi dataset, and one Bengali dataset. The detail description of these datasets are report in Table 1. These datasets are as follows:

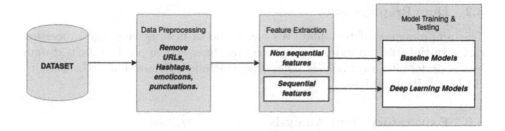

Fig. 1. Methodology pipeline overview

Disaster Tweet Corpus 2020 is a collection of 48 CSV files, each representing a different disaster event. With a total of 150,000 tweets (all files combined), the Disaster Tweet Corpus 2020 dataset provides a substantial volume of data for training and evaluating machine learning and deep learning models.

Kaggle: Natural Language Processing with Disaster Tweets is a total of 10,876 tweets, which were divided into a training data containing 7,613 tweets and a testing data containing 3,263 tweets. For the training set, each row consisted of an ID, the natural language text or tweet, and a corresponding label indicating whether it pertained to a real disaster or not. On the other hand, testing csv file doesnt contain corresponding labels.

Hindi Dataset: The Hindi Disaster Tweet Dataset consists of 20,350 manually created sentences in the Hindi language. Among these sentences, 9,951 are labeled as disaster-related tweets, while the remaining represent non-disastrous or regular tweets. This dataset serves as a valuable resource for training and evaluating machine learning models focused on disaster tweet classification in Hindi.

Bengali Dataset: The Bengali Disaster Tweet Dataset comprises 9,023 manually created tweets in Bengali. Out of these, 5,000 tweets are labeled as disaster-related, while the rest are non-disastrous. This dataset is valuable for training and evaluating machine learning models specifically designed for disaster tweet classification in Bengali.

Table 1. Dataset Description

DATA SETS	Class 0	Class 1	Total
DTC [2]	77728	77734	155465
Kaggles [1]	4342	3271	7613
Hindi Dataset	10399	9951	20350
Bengali Dataset	4023	5000	9023

3.2   Data Preprocessing

The preprocessing of text plays a crucial role in natural language processing tasks. In this study, a series of preprocessing steps were applied to the text data before performing any analysis. These steps included: Removal of URLs, Emojis, Hashtags, Punctuations.

3.3   Exploratory Data Analysis

It is an approach focused on investigating and summarizing key characteristics of the data to gain insights and understand underlying patterns, relationships, and anomalies. Findings in our datasets are discussed below:

Fig. 2. Wordcloud of most occurring words after removing stopwords using DTC [2]

- Most Frequent top-20 words used in DTC dataset tweet text: **the, to, in, of, a, and, for, I, is, you, on, my, this, at, with, it, that, are, from, hurricane** and after removing stopwords: **hurricane, earthquake, people, nepal, flood, im, help, new, like, rt, tornado, get, via, one, irma, us, news, sandy, love, dont.**
- We have obtained a Wordcloud of most occurring words in tweets after removing the stop words shown in Fig. 2.
- Most Frequent top-20 words used in tweet text after removing stopwords in Kaggle dataset: **fire, news, disaster, via, california, suicide, police, people, u, killed, like, hiroshima, pm, storm, fires, crash, families, train, emergency, bomb.**
- After eliminating stop words from a Kaggle dataset, we have generated a Wordcloud representing the most frequent words found shown in Fig. 3.
- Most Frequent top-20 words used in Hindi & Bengali tweet text after stopwords removal is shown in Fig. 6.
- After eliminating stop words from both Hindi & Bengali dataset, we have generated a Wordcloud representing the most frequent words found in these datasets shown in Fig. 4 and Fig. 5.

Fig. 3. Wordcloud of most occurring words after removing stopwords using Kaggle [1]

Fig. 4. Wordcloud of most occurring words after removing stopwords using Hindi dataset

Fig. 5. Wordcloud of most occurring words after removing stopwords using Bengali dataset

'है':8216, 'हैं':3121, 'लोगों':2247, 'कारण':1323, 'भूकंप':1157, 'सुरक्षा':1065, 'प्रभाव':1055, 'प्रभावित':924, 'बाढ़':910, 'आपदा':902, 'था':847, 'जाता':758, 'समय':746, 'तूफान':690, 'किया':656, 'उन्होंने':625, 'खेल':624, 'लोग':601, 'कैसी':591, 'आग':568

'হয়':774, 'গ্যাস':579, 'ফলে':576, 'ভোপাল':563, 'হয়েছে':549, 'প্রাকৃতিক':543, 'হয়েছিলা':528, 'দুর্ঘটনা':500, 'করে':496, 'বন্যার':448, 'গেছে।':442, 'বিভিন্ন':440, 'মানবজনিত':419, 'পরিণামে':401, 'পরিবেশ':379, 'কারণে':370, 'মানুষের':367, 'হয়।':354, 'সৃষ্টি':353, 'পারে':328

Fig. 6. Twenty most occuring words in Hindi and Bengali Dataset

4   Training and Testing

First step is Feature extraction. A model do not understand languages but vectors. To convert the text into vectors we use this technique. After feature extraction, the next step is to train our model. This involves splitting the dataset into two parts- a training set and a testing set. The dataset is divided in a 70:30 ratio, with 70% of the data allocated for training and the remaining 30% reserved for testing. The training set is used to train the model by providing it with input features and their corresponding labels. The model learns from this data by adjusting its internal parameters to capture patterns and relationships. Once the model is trained, it is evaluated using the testing set, which contains unseen data.

4.1   Feature Extraction

In our study, we employ both non-sequential feature extraction and sequential feature extraction techniques. Non-sequential feature extraction is utilized in our baseline models, where the order or sequence of the data points is not considered. We use methods like TF-IDF and CountVectorizer to capture independent characteristics from the data. Both of these techniques are language independent. On the other hand, sequential feature extraction is employed in our deep learning models, where the order and temporal relationships of the data points are important. This approach allows us to capture patterns and trends using techniques like recurrent neural networks (RNNs) or convolutional neural networks (CNNs). By incorporating both non-sequential and sequential feature extraction, we aim to leverage the strengths of each technique and enhance the performance of our models.

4.2   Training

During model training, we employ several machine learning algorithms to identify the best performer among them. We evaluate seven traditional ML algorithms, including Naive Bayes (NB), Logistic Regression (LR), Random Forest (RF), Support Vector Machines (SVM), K-Nearest Neighbors (KNN), Gradient Boosting (GB), Decision Tree (DT) and few DL models named as: LSTM, BiLSTM, BiLSTM with CNN, BERT and DistilBERT. These architectural models like LSTM, BiLSTMs are effective in capturing sequential dependencies in text data. To incorporate local patterns and features, we combined the BiLSTM with Convolutional Neural Networks (CNNs). Moreover, we leveraged the power of pre-trained language models such as BERT and DistilBERT, which are trained on large corpora and can understand the context and semantics of disaster-related text effectively. These algorithms have been finetuned using hyperparameter tuning. Hyperparameter tuning algorithm is given below in Algorithm 1. After finetuning and training the deep learning models, we can move to testing.

Algorithm 1. Hyperparameter Tuning

1: Initialize the hyperparameter search space
2: Split data into training and testing sets
3: Set the best_score to a low value
4: **for** each hyperparameter configuration **do**
5:    Train a model using the current hyperparameters
6:    Evaluate the model on the test set
7:    if test score is better than best_score **then**
8:        Update best_score with the new test score
9:        Update best_hyperparameters with the current hyperparameters
10:    **end if**
11: **end for**
12: Train a final model using the best_hyperparameters

4.3 Testing

For testing we used 30% of data from our dataset. After conducting the testing on all four datasets, we find that SVM performs exceptionally well, exhibiting high accuracy and precision. On the other hand, KNN performs relatively poorly, displaying lower precision compared to the other algorithms as shown in Table 2.

Table 2. Precision Performance of Traditional Machine Learning Classifiers

DATA SETS	NB	LR	RF	SVM	KNN	GB	DT
DTC (Using Count Vec)	90.48	**96.81**	95.34	96.13	95.68	95.56	94.69
DTC (Using Tf_idf)	90.68	94.75	94.11	**95.59**	76.68	89.48	93.35
Kaggle (Using Count Vec)	79.55	79.28	79.00	**80.29**	68.21	72.63	76.24
Kaggle (Using Tf_idf)	79.61	79.92	78.45	**80.14**	66.53	72.63	74.04
Hindi Dataset (Count Vec)	89.91	91.97	91.75	**91.98**	86.16	87.76	88.55
Hindi Dataset (Tf_idf)	89.81	91.25	91.74	**92.07**	85.46	87.51	88.59
Bengali Datset (Count Vec)	94.15	95.82	**96.72**	96.01	91.98	93.12	93.61
Bengali Datset (Tf_idf)	93.42	95.31	**96.08**	96.05	88.09	93.15	93.34

In our research, LSTM outperforms SVM due to its powerful architecture, which includes layers capable of capturing and storing contextual information. LSTM outperforms SVM on various datasets, including the DTC dataset, with an precision of 95.72% and a loss of 0.24. The Bidirectional LSTM (BiLSTM) model outperforms the LSTM model by utilizing the bidirectional nature of the LSTM layers. With the addition of two LSTM layers, it achieves higher precision compared to the LSTM model. Specifically, the BiLSTM achieves impressive precision of 96.18% on the DTC dataset, 97.6% on the Hindi dataset, and 99.3% on the Bengali dataset. However, on the Kaggle dataset, the BiLSTM's precision slightly decreases to 80.11%, which is slightly lower than the 80.26% accuracy achieved by the LSTM model.

Next, we decided to utilize the BiLSTM with CNN architecture due to its ability to capture both local and global dependencies in the data. The architecture of our model includes an embedding layer, a bidirectional LSTM layer, a 1D convolutional layer, a MaxPooling layer, a Flatten layer, and a Dense layer. This model is trained for 10 epochs and the results obtained is mentioned in Table 3.

Table 3. Precision Performance of Deep Learning Classifiers

DATA SETS	LSTM	BiLSTM	BiLSTM +CNN	BERT	DistilBERT
DTC	95.72	96.18	96.75	**97.35**	97.11
Kaggle	80.26	80.11	80.35	83.14	**83.5**
Hindi Dataset	97.1	97.6	97.8	**98.9**	98.8
Bengali Dataset	99.3	99.3	99.1	**99.7**	99.5

After evaluating the performance of other algorithms, we shifted our focus to Transformers-based models, specifically BERT and DistilBERT. These models have gained significant attention and achieved state-of-the-art results in various natural language processing tasks. In our research, we aimed to harness the power of these models for classifying disaster-related tweets accurately. By finetuning these models on a labeled dataset of disaster-related tweets, we aimed to train them to accurately classify new tweets into relevant categories, distinguishing between disaster-related and non-disaster tweets. The finetuning algorithm for this task is described in Algorithm 1. The hyperparameters, including batch size, learning rate, maximum sequence length, and epochs, were systematically tuned to explore different combinations and determine the optimal settings that yield the best results.

Table 4. Comparison with other research

Research	Models	Score
Chanda et al. [5]	Logistic Regression	72.93
	BERT-BiLSTM	83.08
Pratama et al. [20]	CNN-GRU	83.3
Our Research	SVM	80.29
	BERT	83.144
	DistilBERT	83.5

When comparing our study results with existing research, we find that our findings align with the previous studies in several aspects. However, there are also some notable differences and novel insights that our study brings to the table. Comparison of our research with existing research is noted in Table 4. The scores used for comparison were obtained by submitting the predicted labels of the test.csv dataset from Kaggle [1]. These scores were calculated by the Kaggle website itself. Our research showcased the outstanding performance and impressive results of both BERT and DistilBERT, which were achieved through the implementation of hyperparameter optimization techniques.

After analyzing the achieved outcomes, we compiled a dataset consisting of tweets in English, Hindi, and Bengali, combining all three languages. This dataset was then fed into our BERT model, which yielded an accuracy of 98.28% and a loss of approximately 0.0718.

5   Results

The results of our evaluation, presented in Table 2, provide a comprehensive analysis of seven machine learning algorithms using two vectorizer techniques: TF-IDF and Count Vectorizer. The performance of these algorithms was assessed across four datasets: DTC [2], Kaggle [1], Hindi, and Bengali.

Our analysis revealed that the algorithms employing the Count Vectorizer consistently outperformed those using TF-IDF across all datasets. This suggests that the frequency-based approach of Count Vectorizer yielded superior results in classifying disaster-related tweets. Among the seven ML algorithms, Support Vector Machine (SVM) exhibited the highest overall performance across all datasets except for Bengali. SVM consistently achieved higher accuracy, precision, recall, and F1-score compared to the other algorithms in the English and Hindi datasets. However, in the Bengali dataset, another algorithm demonstrated superior performance. While the exact ranking and performance varied across datasets, the general trend of SVM outperforming other algorithms persisted.

In addition to the ML algorithms, we also employed deep learning models to classify disaster-related tweets in English, Hindi, and Bengali datasets. Among the evaluated deep learning models, including LSTM, BiLSTM, BiLSTM with CNN, BERT, and DistilBERT, BERT and DistilBERT consistently showcased outstanding performance across all datasets as shown in Table 3. These models consistently outperformed the other deep learning techniques and achieved notable accuracy in accurately classifying disaster-related tweets.

Although the ML algorithms demonstrated competitive performance, they were generally surpassed by BERT and DistilBERT in most cases. However, it is important to note that the ML algorithms still achieved respectable results. Overall, our research highlights the exceptional effectiveness of BERT and DistilBERT as powerful deep learning models for disaster-related tweet classification. These models offer valuable insights for tracking and responding to disaster events, showcasing remarkable performance and reliability across diverse language contexts.

6   Conclusion

The study involved an extensive evaluation of seven machine learning algorithms using two vectorizer techniques, TF-IDF and Count Vectorizer, across four datasets: DTC, Kaggle, Hindi, and Bengali. The findings consistently revealed that algorithms utilizing Count Vectorizer consistently outperformed those employing TF-IDF, highlighting the effectiveness of the frequency-based approach in classifying disaster-related tweets. Support Vector Machine emerged as the top-performing ML algorithm across most datasets, except for Bengali, where another algorithm exhibited superior results. Notably, deep learning models, specifically BERT and DistilBERT, consistently demonstrated exceptional performance across all datasets, surpassing both the ML algorithms and other deep learning techniques. Based on these findings, a dataset comprising tweets in all three languages, i.e., English, Hindi, and Bengali was compiled and fed into the BERT model which resulting in an accuracy of 98.28% and a loss of approximately 0.0718. BERT and DistilBERT proved to be powerful models for classifying disaster tweets, delivering exceptional accuracy and reliability across diverse language contexts.

References

1. Covid-19 tweet classification challenge. https://www.kaggle.com/competitions/nlp-getting-started
2. Dtc 2020. https://zenodo.org/record/3713920
3. Alhammadi, H.: Using machine learning in disaster tweets classification (2022)
4. Arora, S., Kumar, S., Kumar, S.: Artificial intelligence in disaster management: a survey. In: Saraswat, M., Chowdhury, C., Kumar Mandal, C., Gandomi, A.H. (eds.) ICDSA 2022, vol. 2, pp. 793–805. Springer, Cham (2023). https://doi.org/10.1007/978-981-19-6634-7_56
5. Chanda, A.K.: Efficacy of BERT embeddings on predicting disaster from twitter data. arXiv preprint arXiv:2108.10698 (2021)
6. Deb, S., Chanda, A.K.: Comparative analysis of contextual and context-free embeddings in disaster prediction from twitter data. Mach. Learn. Appl. **7**, 100253 (2022)
7. Dharma, L.S.A., Winarko, E.: Classifying natural disaster tweet using a convolutional neural network and BERT embedding. In: 2022 2nd International Conference on Information Technology and Education (ICIT&E), pp. 23–30 (2022). https://doi.org/10.1109/ICITE54466.2022.9759860
8. Dharma, L.S.A., Winarko, E.: Classifying natural disaster tweet using a convolutional neural network and BERT embedding. In: 2022 2nd International Conference on Information Technology and Education (ICIT&E), pp. 23–30. IEEE (2022)
9. Fatyanosa, T.N., Aritsugi, M.: Effects of the number of hyperparameters on the performance of GA-CNN. In: 2020 IEEE/ACM International Conference on Big Data Computing, Applications and Technologies (BDCAT), pp. 144–153 (2020). https://doi.org/10.1109/BDCAT50828.2020.00016
10. Gulati, N., Agarwal, A., Aggarwal, A., Bhutani, N., Kapur, R.: Ensembled multi-detector aggregation for disaster detection (EMAD). In: 2023 13th International Conference on Cloud Computing, Data Science & Engineering (Confluence), pp. 593–596 (2023). https://doi.org/10.1109/Confluence56041.2023.10048857

11. Kanimozhi, T., Belina, V.J., Sara, S.: Classification of tweet on disaster management using random forest. In: Rajagopal, S., Faruki, P., Popat, K. (eds.) ASCIS 2022, Part I, pp. 180–193. Springer, Cham (2023)

12. Koshy, R., Elango, S.: Multimodal tweet classification in disaster response systems using transformer-based bidirectional attention model. Neural Comput. Appl. **35**, 1607–1627 (2022)

13. Kumar, A., Singh, J.P., Saumya, S.: A comparative analysis of machine learning techniques for disaster-related tweet classification. In: 2019 IEEE R10 Humanitarian Technology Conference (R10-HTC)(47129), pp. 222–227. IEEE (2019)

14. Lamsal, R., Kumar, T.V.: Twitter-based disaster response using recurrent nets. In: Research Anthology on Managing Crisis and Risk Communications, pp. 613–632. IGI Global (2023)

15. Le, A.D.: Disaster tweets classification using bert-based language model. arXiv preprint arXiv:2202.00795 (2022)

16. Madichetty, S., Muthukumarasamy, S.: Detection of situational information from twitter during disaster using deep learning models. Sādhanā **45**(1), 1–13 (2020)

17. Madichetty, S., Muthukumarasamy, S., Jayadev, P.: Multimodal classification of twitter data during disasters for humanitarian response. J. Ambient. Intell. Humaniz. Comput. **12**(11), 10223–10237 (2021)

18. Ningsih, A., Hadiana, A.: Disaster tweets classification in disaster response using bidirectional encoder representations from transformer (BERT). In: IOP Conference Series: Materials Science and Engineering, vol. 1115, p. 012032. IOP Publishing (2021)

19. Prasad, R., Udeme, A.U., Misra, S., Bisallah, H.: Identification and classification of transportation disaster tweets using improved bidirectional encoder representations from transformers. Int. J. Inf. Manag. Data Insights **3**(1), 100154 (2023)

20. Pratama, R.A., Pardede, H.F.: Disaster tweet classifications using hybrid convolutional layers and gated recurrent unit. Int. J. Comput. Dig. Syst. **13**(1), 1–1 (2023)

21. Rathod, J., Rathod, G., Upadhyay, P., Vakhare, P.: Disaster tweet classification using ml. In: 2022 International Conference on Applied Artificial Intelligence and Computing (ICAAIC), pp. 523–527. IEEE (2022)

22. Ritchie, H., Rosado, P., Roser, M.: Natural disasters. Our World in Data (2022). https://ourworldindata.org/natural-disasters

23. Sirbu, I., Sosea, T., Caragea, C., Caragea, D., Rebedea, T.: Multimodal semi-supervised learning for disaster tweet classification. In: Proceedings of the 29th International Conference on Computational Linguistics, pp. 2711–2723 (2022)

24. Snyder, L.S., Lin, Y.S., Karimzadeh, M., Goldwasser, D., Ebert, D.S.: Interactive learning for identifying relevant tweets to support real-time situational awareness. IEEE Trans. Visual Comput. Graphics **26**(1), 558–568 (2020). https://doi.org/10.1109/TVCG.2019.2934614

25. Song, G., Huang, D.: A sentiment-aware contextual model for real-time disaster prediction using twitter data. Future Internet **13**(7), 163 (2021)

26. Toraman, C., Kucukkaya, I.E., Ozcelik, O., Sahin, U.: Tweets under the rubble: detection of messages calling for help in earthquake disaster. arXiv preprint arXiv:2302.13403 (2023)

Enhancing Mitotic Cell Segmentation: A Transformer Based U-Net Approach

Anusree Kanadath[✉], J. Angel Arul Jothi, and Siddhaling Urolagin

Department of Computer Science, Birla Institute of Technology & Science, Pilani,
Dubai Campus, Dubai International Academic City, Dubai, United Arab Emirates
{p20180904,angeljothi,siddhaling}@dubai.bits-pilani.ac.in

Abstract. Mitosis segmentation plays a vital role in early cancer detection, facilitating the accurate identification of dividing cells in histopathology images. Manual mitosis counting is time-consuming and subjective, prompting the need for automated approaches to improve efficiency and accuracy. In this study, we have developed a transformer-based U-Net model that combines the effectiveness of transformers which were originally designed for natural language processing (NLP) tasks, with the efficiency of the U-Net architecture to effectively capture both high-level and low-level features in histopathology images. We train and evaluate the model on the GZMH dataset and compare its performance against other deep models such as U-Net, U-Net++ and Mobilenetv2-based U-Net. The results demonstrate that transformer-based U-Net model is better in terms of accuracy, recall, precision, F1-score and Dice coefficient. This study represents a significant advancement in mitosis segmentation, contributing to improved cancer detection and prognosis.

Keywords: Histopathology · Image segmentation · Deep learning · Encoder-decoder model · U-Net model · Vision Transformers

1 Introduction

According to the World Health Organization (WHO), breast cancer stands as a significant global health concern, with over 2.3 million new cases each year, making it the most prevalent cancer among adults. [1]. Breast cancer is the second most significant cause of cancer-related mortality in women. According to a study conducted in 2020, approximately one million children have lost a parent to cancer, with breast cancer accounting for 25% of these deaths. To address this, the WHO has released the Global Breast Cancer Initiative (GBCI) Framework, which aims to save 2.5 million lives by 2040. The framework emphasizes health promotion for early breast cancer detection, prompt diagnosis and comprehensive management [2].

Early cancer detection improves treatment options and survival rates. This reduces the risk of disease progression. It makes it possible to create treatment plans that are unique to each type of cancer as well as the patient, which helps improve therapeutic results while simultaneously lowering the risk of adverse effects.

R. Muthalagu et al. (Eds.): CINS 2023, CCIS 1978, pp. 130–142, 2024.
https://doi.org/10.1007/978-3-031-48984-6_11

Understanding mitotic cell activity is essential for the early detection and treatment of cancer. Mitotic cell activity is a key indication of cancer aggressiveness and prognosis. Cancers with a high mitotic cell count frequently exhibit rapid tumor growth and aggressiveness. Mitotic activity is high in breast cancer, melanoma, non-small cell lung cancer, pancreatic ductal adenocarcinoma, some uterine cancer subtypes and aggressive bladder cancer. Evaluating mitotic cell activity along with other clinical and genetic characteristics is crucial for cancer diagnosis, prognosis and patient-specific treatment.

Mitosis segmentation also known as mitotic cell segmentation, is a key aspect of digital pathology. It involves accurately identifying and separating mitotic cells in whole slide images (WSIs) [3]. Mitotic cell segmentation focuses on identifying cells undergoing mitosis, which is an essential phase of the cell cycle in which a single cell divides into two identical daughter cells. Mitosis segmentation from Hematoxylin and Eosin (H&E) stained histopathological images is challenging due to the varying appearances during the four mitosis stages (prophase, metaphase, anaphase and telophase), low occurrence of mitotic compared to non-mitotic cells, and similarities with other cell types (apoptotic cells, dense nuclei). Automated mitosis segmentation is a cutting-edge method that uses computer vision and machine learning to automatically find and separate mitotic cells in images. In the past, identifying mitotic cells manually was a time-consuming and error-prone process, but with the advent of advanced algorithms and deep learning models, this process has been improved.

Compared to conventional algorithms, deep learning algorithms provide a more precise region segmentation. The autoencoder, which is a type of artificial neural network, is one such effective model for image segmentation [4]. The autoencoder consists of an encoder part that extracts features from the input image and a decoder part that automatically reconstructs the image from these extracted features. U-Net is a type of autoencoder framework that is designed to solve semantic segmentation problems [5]. A key aspect of the U-Net model is its utilization of skip connections, which share contextual information extracted in the contracting path to the expanding path, enhancing the model's segmentation capabilities.

In recent years, transformers, which were originally developed for natural language processing (NLP) tasks, have also shown remarkable performance in the field of image processing [6]. These models, called vision transformers (ViTs), have made a lot of success in tasks like image recognition, segmentation and generation [7]. Transformers can handle visual data quickly and better than traditional convolutional neural networks (CNNs) on some tasks because they treat images as sequences of patches or tokens. Their self-attention mechanism enables them to evaluate the significance of various patches when analyzing an image, allowing for more accurate recognition and understanding of complex visual patterns.

In this work we apply a transformer based U-Net model for the accurate segmentation of mitosis from histopathology images. By integrating transformers and U-Net architecture, we aim to identify mitotic cells in WSIs with better accuracy and efficacy. The self-attention mechanism in ViTs enables the model to effectively capture high level features from WSIs. On the other hand, the U-Net architecture, with its skip connections further improves the model's ability to capture low level feature maps from WSIs. Using a transformer based U-Net model can help advance the field of mitosis segmentation leading to better cancer diagnosis and prognosis. To the best of our knowledge, this is the first time a transformer based U-Net architecture has been applied specifically to mitosis segmentation task. The main objectives of this study are as follows:

- Develop a transformer based U-Net model to segment mitosis regions from histopathology images
- Apply and analyze the effectiveness of the model in segmenting mitosis regions from breast histopathology images
- Evaluate the performance of transformer based U-Net model against other deep models like U-Net, U-Net++ and MobileNetV2 based U-Net
- Compare the performance of the model with previous works done on the dataset.

The following is the structure of the subsequent sections of the paper. Section 2 provides the literature survey. Section 3 gives the detailed description of the dataset used in this study. In Sect. 4, the model's architecture is explained in detail. Section 5 describes the experimental set-up. Results and discussions are explained in Sect. 6. The final section of the paper summarizes the main findings and implications of the work.

2   Literature Survey

Digital mitosis segmentation has improved over the years as researchers develop new methods for identifying mitotic cells in histopathology images. This literature review analyzes the different approaches given by distinct studies, each of which provides unique insights and solutions to this crucial task.

Anabia Sohail et al. introduced a novel mitosis detection framework named MP-MitDet for histopathological images of breast cancer [8]. Their approach included a label refiner for weak label representation, tissue-level mitotic region selection, deep instance-based detection and segmentation, and cell-level refinement. When tested on the TUPAC16 dataset, the framework achieved a F-score of 0.75, recall of 0.76, precision of 0.71, and area under the precision-recall curve of 0.78. Razavi et al. proposed an automatic method based on a conditional generative adversarial network for segmenting mitosis and nuclei in diverse H&E breast cancer pathology images [9]. Multiple datasets such as TUPAC16, ICPR14 and ICPR12, were utilized to evaluate the accuracy of the proposed method. This approach produced more realistic and accurate segmentation maps than U-Net models.

Li et al. presented an accurate method for detecting and counting mitotic cells on histopathological samples based on a multi-stage deep learning framework [10]. The strategy consisted of a deep segmentation network to generate mitosis regions from weak labels, a deep detection network for accurate mitosis localization, and a deep verification network to eliminate false positives. Wang et al. proposed SCMitosis, a two-stage mitotic segmentation and classification approach [11]. The first stage used the depthwise separable convolution residual block and channel-spatial attention gate to achieve high segmentation performance and recall rate. In the second stage, a classification network was cascaded to improve the efficacy of mitotic cell detection. On the ICPR 2012 dataset, the model had a F-score of 0.86. Zhang et al. used hand-crafted and deep features to segment mitotic cells in WSIs [12]. Four measure indices from the Gray Level Co-occurrence Matrix were used to extract handcrafted features, whereas natural image knowledge transfer was used to obtain deep features. AMIDA13 was used for experiments. The suggested technique outperformed the U-Net model on the test and validation datasets.

According to the literature review, deep learning models have been used to locate mitosis in breast cancer histopathology images. The absence of transformer-based architectures for mitosis segmentation represents a significant research area. Attention mechanisms can be used to enhance segmentation accuracy by integrating transformers into the U-Net framework. Therefore, we are applying a transformer-based U-Net model for mitosis segmentation to enhance the accuracy and interpretability of mitosis segmentation.

3   Dataset Description

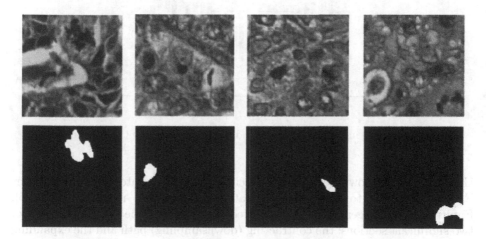

Fig. 1. Basic architecture of U-Net model for image segmentation.

For this study, we used a data set from China's Ganzhou Municipal Hospital (GZMH) [11]. The dataset consists of 55 WSIs from 22 patients that were digitized using a digital section scanner at 40x magnification and 0.25 um/pixel resolution. The mitosis nuclei images are carefully annotated and double-checked by professional pathologists. The training set contains 1832 mitotic nuclei in 1192 High-Power Field (HPF) images, while the testing set contains 523 mitotic nuclei in 342 HPF images from different patients. There are two parts to the dataset: the training set, which has 48 WSIs and the testing set, which has 7 WSIs. Each image in the dataset is a RGB image with a resolution of 2084 × 2084 pixels, along with corresponding binary masks in black and white images. Figure 2 shows the sample images and corresponding masks from the GZMH dataset.

4   Methodology

In this section, the transformer based U-Net model for mitosis segmentation from histopathology images is described in detail. The first two subsections gives some basic information about the U-Net architecture and vision transformers. The following subsection explains the transformer based U-Net model and its workflow.

4.1   U-Net Architecture

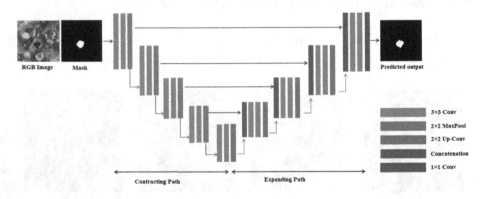

Fig. 2. Sample images and corresponding masks form the GZMH dataset.

U-Net is a well-known network architecture that was created by Ronneberger, Ficher, and Brox to address segmentation challenges in medical images [5]. Figure 1 depicts a general U-Net architecture. The U-shaped design consists of two significant sections: the contracting (downsampling) path and the expanding (upsampling) path. During the training of the model, histopathology images and their corresponding masks are given as input. The contracting path, located on the left side of Fig. 1, is comprised of multiple groups of layers, with each group consisting of a pair of 3 × 3 convolution layers followed by a 2 × 2 max pooling layer. Each layer receives as input the feature maps generated by the previous layer. U-Net effectively captures contextual information about the input image during the downsampling phase, which is then propagated to the upsampling path for further processing. The upsampling path on the right side of Fig. 1 consists of multiple groups of layers, with each group consisting of two 3×3 convolution layers followed by a 2×2 upsampling layer. This upsampling path is essential for recreating the segmented mask based on contextual information acquired during the downsampling phase. Significant to the U-Net model is the presence of horizontal connections, also known as skip connections, which connect higher-level feature maps from the contracting path to the expansion path, thereby improving segmentation accuracy. Layers of concatenation are used to combine feature maps of the same dimension. The final layer in the expansion path is a 1×1 convolutional layer, which maps the previous layer's feature maps to a binary segmented output.

4.2   Vision Transformer Model

Transformers, which were originally designed for NLP applications, have acquired widespread popularity due to their exceptional performance and adaptability in comprehending textual data. The concept of vision transformers extends this method to the image domain by modifying transformers for image processing [7,13]. ViTs divide input images into visual tokens, which subsequently embed the tokens into encoded vectors. The transformer encoder processes these vectors along with the patch positions, creating an architecture similar to NLP transformers [14,15]. Figure 3 depicts a general ViT architecture. The basic steps of ViTs are as follows:

First, the input image is divided into non-overlapping patches of equal size, with the number of patches (n) determined by $n = hw/p^2$, where h, w, and p represent the image's height, width, and patch size, respectively. The patches are then flattened and linearly projected, converting them to a one-dimensional format for further processing. In addition, positional encoding is applied to the flattened patches. Before feeding the sequence of patches into the transformer encoder, an additional classification token is added to the beginning of the sequence in order to make predictions about the image's class.

Self-attention mechanism is a key component of ViTs. In self-attention, the input image patches are processed as a sequence of vectors. These patches undergo a normalization procedure, resulting in a sequence of encoded patches. Three linear layers are applied to encoded patches to generate three distinct projections: query (Q), key (K), and value (V). Attention is calculated by taking the dot product of Q and the transpose of K. In order to reduce the output, it is divided by the square root of the dimension d. This output is then transmitted through a softmax function to generate a weight matrix. This matrix is used to perform second matrix multiplication with V. The equation for the attention mechanism is given in Eq. 1.

$$Attention(Q, K, V) = softmax\left(QK^T/\sqrt{d}\right) V \tag{1}$$

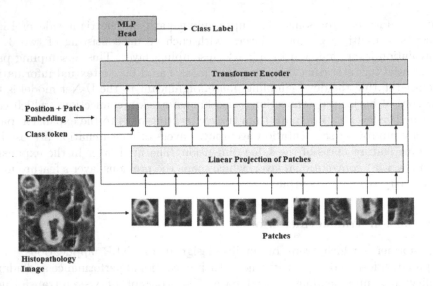

Fig. 3. Basic architecture of ViT model.

The transformer encoder of ViT uses multi-head attention. In multi-head attention, the projections of Q, K, and V are repeated multiple times. Every projection is processed simultaneously with the self-attention mechanism. Finally, the outputs of all attention heads are concatenated. This method permits ViT to extract numerous and specific features from the input image, thereby enhancing its capacity to perform image classification tasks effectively.

4.3   Vision Transformer Based U-Net Model

In this study, we developed a transformer based U-Net model inspired by the TransUnet model [16]. We have applied the model for segmenting mitosis regions in H&E stained breast histopathology images. Figure 4 depicts a segmentation model, illustrating the incorporation of vision transformers to improve the segmentation performance of the U-Net architecture.

This architecture employs a hybrid encoder that combines CNN and transformer models. Transformer-only encoders may have limited localization capabilities, resulting in the loss of some features. This limitation arises because transformers are primarily designed to capture high-level features rather than low-level features. Consequently, a CNN is employed to extract low-level features from input images, serving as the backbone for feature extraction. The hidden features extracted by the CNN are then linearly projected and passed into the transformer unit, which consists of 12 transformer layers. Each transformer layer comprises a sequence of layers, including a normalization layer, multi-head attention layer, another normalization layer, and a multi-layer perception layer.

The decoding process is similar to the decoding process in the U-Net architecture. Initially, the output from the transformer unit is subjected to a 3×3

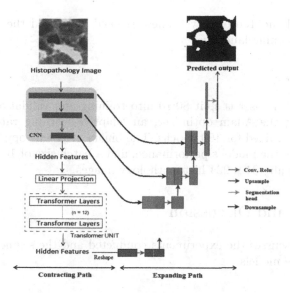

Fig. 4. Architecture of the segmentation model.

convolutional layer with rectified linear unit (ReLU) activation, followed by upsampling. The result is then concatenated with the output of the third-level CNN feature extractor. Subsequently, the resulting feature maps are processed through another 3 × 3 convolution layer with ReLU activation, followed by further upsampling and concatenation with the output from the second-level CNN feature extractor. This process is repeated again. Finally, the output from the decoder is directed to the segmentation head, where the segmented image corresponding to the original input image is generated.

5 Experimental Setup

5.1 Data Preprocessing

Each image in the training set has been split into 224 × 224-pixel patches, to increase the dataset size and diversity. After patching and preprocessing, the training dataset has 2140 images with 224 × 224 masks and the testing set has 610 224 × 224-pixel images and masks. Data augmentation techniques are performed to each training set image to enhance training data. These techniques include image rotation with angles of 90, 180, and 270°, as well as horizontal and vertical flips. By applying these augmentation methods, the final train set comprises a total of 12840 images.

5.2 Hardware Software Setup

The training process was conducted on a computer with an Intel(R) Xeon(R) W-2123 CPU, 16 GB of RAM, and a 1 TB hard drive. For implementing the model, we used the TensorFlow framework, and developed the code using the Python programming language.

5.3   Training

The augmented dataset is split 80:10 into training and validation sets to train a model. Using the Adam optimizer, an adaptive learning rate optimization algorithm, it is trained for 100 epochs. The binary cross-entropy loss function is used to evaluate the model's performance, and a batch size of 16 is employed to efficiently train all 224×224 histopathology images.

6   Results and Discussion

This section discusses the experiments conducted and the segmentation results obtained by the models.

6.1   Experiments Conducted

In this study, three different segmentation models for histopathology images have been developed to allow for a fair comparison with transformer based U-Net. These models include:

- U-Net: The U-Net model is the basic architecture, made up of an encoder and decoder path connected by skip connections [5].
- U-Net++: The U-Net++ model is an enhanced version of U-Net in which direct skip connections are replaced with dense skip connections inspired by the DenseNet architecture [17].
- MobileNetV2-based U-Net: In the MobileNetV2-based U-Net, we substituted the encoder portion of the U-Net with a pretrained MobileNetV2 model on ImageNet [18]. MobileNetV2 is known for its lightweight and effective architecture, which may result in quicker training and inference [19].

To ensure a fair comparison of these models, we have evaluated their performance on the same dataset, using same evaluation metrics, and follow a rigorous training and validation process.

6.2   Segmentation Results

For each model, the accuracy, precision, recall, F1-score, and dice coefficient were recorded. The segmentation results are shown in Table 1. For every metric, the best value has been highlighted.

Table 1. Segmentation results of the models.

Metrics	U-Net	U-Net++	Mobilenetv2-U-Net	Transformer based U-Net
Dice Coefficient	0.47	0.57	0.66	**0.83**
Accuracy	0.96	0.97	0.97	**0.98**
Recall	0.31	0.36	0.45	**0.60**
Precision	0.91	0.82	0.92	**0.98**
F1-score	0.46	0.50	0.60	**0.74**

From the table, it is observed that the transformer-based U-Net model achieved the highest Dice Coefficient of 0.83, indicating that it performed better in terms of overall similarity between the predicted and ground truth segmentations. Mobilenetv2-U-Net also performed well with a Dice Coefficient of 0.66, while U-Net and U-Net++ achieved lower scores of 0.47 and 0.57, respectively. The transformer-based U-Net model demonstrated the highest accuracy of 0.98, outperforming the other models. U-Net, U-Net++ and Mobilenetv2-U-Net had slightly lower accuracy scores of 0.96, 0.97 and 0.97, respectively. In addition, the transformer-based U-Net model had a highest recall of 0.60 and a highest precision of 0.98. Overall, the transformer-based U-Net model outperformed the U-Net, U-Net++, and Mobilenetv2-U-Net models across all evaluation metrics, with an F1-score of 0.74, demonstrating its efficacy in mitotic segmentation.

This result implies that the transformer-based U-Net model consistently outperforms other CNN-based deep models, such as U-Net and U-Net++. Despite the use of a large number of deep convolutional blocks by U-Net and U-Net++, the transformer-based architecture achieves superior segmentation results. In addition, the transformer-based model outperforms the MobileNetV2-based U-Net model which is a pretrained model on ImageNet dataset. This demonstrates the effectiveness and potential of transformer-based models for mitosis segmentation in histopathology images, offering promising advancements in the field of medical image analysis.

6.3   Comparison with Previous Work Done on the GZMH Dataset

We collected and tabulated previous works on the GZMH dataset in order to have a fare comparison with transformer based U-Net model. Table 2 shows the previous works and the precision, recall and F1-scores obtained for the GZMH dataset. SCMitosis achieved the highest recall among the listed models, 0.73, indicating its proficiency in accurate mitotic cell segmentation. However, it is inferior to the transformer-based U-Net model in terms of precision and F1-score. On the other hand, U-Net, SegNet, R2U-Net, LinkNet34, and DeepLabV3+ exhibited relatively higher recall values. However, these models compromised on precision and F1-score when compared to transformer-based U-Net model. The transformer-based U-Net model segment mitotic cells accurately than other models. It also has a good F1 score, balancing precision and recall.

Table 2. Previous results on the GZMH dataset.

Model	Precision	Recall	F1-score
U-Net [11]	0.3022	0.7784	0.4353
SegNet [11]	0.2904	0.8304	0.4304
R2U-Net [11]	0.3164	0.8054	0.4543
LinkNet34 [11]	0.3623	0.7225	0.4826
DeepLabV3+ [11]	0.2937	0.7900	0.4282
SCMitosis [11]	0.4278	0.7325	0.5402
Transformer based U-Net	0.98	0.60	0.74

On the GZMH dataset, the model performed well, demonstrating its potential to improve mitotic segmentation and cancer detection and prognosis. Figure 5 shows three sample images from the test set, along with the corresponding ground truth mask and the segmentation output of the transformer-based U-Net model, respectively.

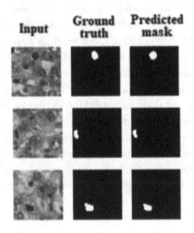

Fig. 5. Segmentation mask generated by transformer based U-Net model.

7   Conclusion

In this work, we have developed a transformer-based U-Net model for segmenting mitotic regions in breast histopathology images. By replacing the encoder with a CNN followed by transformer blocks, we enhance the model's ability to capture long-range dependencies. We trained the model on an augmented GZMH dataset and compared its performance with other segmentation models like U-Net, U-Net++, and mobilenetv2 based U-Net on the same dataset. Our results demonstrate that our proposed model outperforms these state-of-the-art algorithms, offering a promising solution for accurate and efficient mitotic region segmentation in breast cancer diagnosis. To enhance practical applicability, future research should encompass diverse histopathology datasets and imaging conditions to assess the model's robustness and generalizability comprehensively.

References

1. WHO launches new roadmap on breast cancer (2023). https://www.who.int/. Accessed 28 Jul 2023
2. Global breast cancer initiative implementation framework: assessing, strengthening and scaling up of services for the early detection and management of breast cancer: executive summary (2023). https://www.who.int/publications/i/item/9789240067134. Accessed 28 July 2023
3. Dominguez-Brauer, C., Thu, K.L., Mason, J.M., Blaser, H., Bray, M.R., Mak, T.W.: Targeting mitosis in cancer: emerging strategies. Mol. Cell **60**(4), 524–536 (2015)
4. Bank, D., Koenigstein, N., Giryes, R.: Autoencoders., Deep Learning in Science (2021)
5. Ronneberger, O., Fischer, P., Brox, T.: U-net: convolutional networks for biomedical image segmentation. In: Navab, N., Hornegger, J., Wells, W.M., Frangi, A.F. (eds.) MICCAI 2015. LNCS, vol. 9351, pp. 234–241. Springer, Cham (2015). https://doi.org/10.1007/978-3-319-24574-4_28
6. Vaswani, A., et al.: Attention is all you need. In: Guyon, I., et al. (eds.) Advances in Neural Information Processing Systems, vol. 30. Curran Associates Inc. (2017)
7. Dosovitskiy, A., et al.: An image is worth 16x16 words: transformers for image recognition at scale. In: 9th International Conference on Learning Representations, ICLR 2021, Virtual Event, Austria, May 3–7, 2021, OpenReview.net (2021)
8. Sohail, A., Khan, A., Wahab, N., Zameer, A., Khan, S.: A multi-phase deep CNN based mitosis detection framework for breast cancer histopathological images. Sci. Rep. **11**, 6215 (2021)
9. Razavi, S., Khameneh, F.D., Nouri, H., Androutsos, D., Done, S.J., Khademi, A.: Minugan: dual segmentation of mitoses and nuclei using conditional GANs on multi-center breast H&E images. J. Pathol. Inform. **13**, 100002 (2022)
10. Li, C., Wang, X., Liu, W., Latecki, L.J.: Deepmitosis: mitosis detection via deep detection, verication and segmentation networks. Med. Image Anal. **45**, 01 (2018)
11. Wang, H., et al.: A novel dataset and a deep learning method for mitosis nuclei segmentation and classification, ArXiv:abs/2212.13401 (2022)
12. Zhang, Y., Chen, J., Pan, X.: Multi-feature fusion of deep networks for mitosis segmentation in histological images. Int. J. Imaging Syst. Technol. **31**, 09 (2020)
13. Ruan, B., Shuai, H.-H., Cheng, W.-H.: Vision transformers: state of the art and research challenges. ArXiv:abs/2207.03041 (2022)
14. Bai, Y., Mei, J., Yuille, A., Xie, C.: Are transformers more robust than CNNs? In: Advances in Neural Information Processing Systems (2021)
15. Raghu, M., Unterthiner, T., Kornblith, S., Zhang, C., Dosovitskiy, A.: Do vision transformers see like convolutional neural networks? In: Neural Information Processing Systems (2021)
16. Chen, J., et al.: TransuNet: transformers make strong encoders for medical image segmentation. ArXiv:abs/2102.04306 (2021)

17. Zhou, Z., Rahman Siddiquee, M.M., Tajbakhsh, N., Liang, J.: UNet++: a nested U-net architecture for medical image segmentation. In: Stoyanov, D., et al. (eds.) DLMIA/ML-CDS -2018. LNCS, vol. 11045, pp. 3–11. Springer, Cham (2018). https://doi.org/10.1007/978-3-030-00889-5_1
18. Kanadath, A., Jothi, J.A.A., Urolagin, S.: Histopathology image segmentation using mobilenetv2 based U-net model. In: 2021 International Conference on Intelligent Technologies (CONIT), pp. 1–8 (2021)
19. Sandler, M., Howard, A.G., Zhu, M., Zhmoginov, A., Chen, L.-C.: Mobilenetv2: inverted residuals and linear bottlenecks. In: 2018 IEEE/CVF Conference on Computer Vision and Pattern Recognition, pp. 4510–4520 (2018)

Author Index

Printed in the United States
by Baker & Taylor Publisher Services